Previous books by the same authors include:

BY JOHN GRIBBIN

Climatic Change (ed.)
The Death of the Sun
Forecasts, Famines and Freezes
Future Worlds
Galaxy Formation: *A Personal View*
Genesis: *The Origins of Man and Universe*
Our Changing Climate
Timewarps
Weather Force: *Climate and Its Impact on Our World*
White Holes: *Cosmic Gushers in the Universe*

Novel:
The Sixth Winter (with Douglas Orgill)

BY JEREMY CHERFAS

Not Work Alone:
A Cross-Cultural View of Activities Superfluous to Survival
(ed., with Roger Lewin)

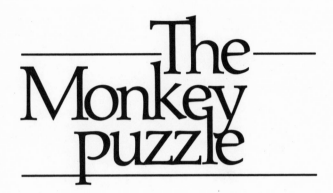

The Monkey puzzle

The Monkey puzzle

RESHAPING THE EVOLUTIONARY TREE

John Gribbin & Jeremy Cherfas

Pantheon Books
New York

Library of Congress Cataloging in Publication Data

Gribbin, John R.
The monkey puzzle.

Originally published: London: Bodley Head, 1982.
Bibliography: p.
Includes index.
1. Human evolution.   I. Cherfas, Jeremy.
II. Title.
GN281.G74   1982   573.2   82-47889
ISBN 0-394-52794-1   AACR2

He who wishes to decide whether man is the modified descendant of some pre-existing form, would probably first enquire whether man varies, however slightly, in bodily structure and in mental faculties; and if so, whether the variations are transmitted to his offspring in accordance with the laws which prevail with the lower animals. Again, are the variations the result, as far as our ignorance permits us to judge, of the same general causes, and are they governed by the same general laws, as in the case of other organisms? . . . Is man subject to similar malconformations, the result of arrested development, of reduplication of parts &, and does he display in any of his anomalies reversion to some former and ancient type of structure?

Charles Darwin

I know my molecules had ancestors, the paleontologist can only hope that his fossils had descendants.

Vincent Sarich

GEOLOGICAL ERA	GEOLOGICAL PERIOD	EPOCH
Present	Quaternary 3 My BP	Recent 11 000 y Before Present
Cainozoic (about 1·5 percent of Earth history)	Tertiary 65 My BP	Pleistocene 3 My BP
70 MY		Pliocene 7 My BP
	Cretaceous 135 My BP	Miocene 25 My BP
Mesozoic (about 3 percent of Earth history)	Jurassic 190 My BP	Oligocene 40 My BP
		Eocene 60 My BP
	Triassic 225 My BP	Palaeocene 70 My BP
230 MY	Permian 285 My BP	
	Carboniferous 350 My BP	
Palaeozoic (about 8 percent of Earth history)	Devonian 400 My BP	
	Silurian 440 My BP	
	Ordovician 505 My BP	
about 570 MY	Cambrian 570 My BY	
Pre-Cambrian (about 90 percent of Earth history)		

Chart of Geological Time
My = million years
BP = before present

CONTENTS

For
Sylvia and Min

PREFACE

This book grew out of the bewilderment of one of the authors, who has no formal training in evolutionary science, at the lack of response from those who have had such training to the fascinating work of Vincent Sarich and Allan Wilson, of Berkeley in California, and their colleagues. The work, published in widely read and respectable scientific journals such as *Nature* and *Science,* is about the origin and evolution of all living things. That means it is about mankind. By studying the molecules of life from living species, rather than the preserved fossil remains of extinct ones, Sarich and Wilson showed, as long ago as 1967, that our kinship to the African apes is much closer than any mainstream palaeontologist has ever acknowledged. To an astronomer with an outsider's interest in the subject, the evidence seemed dramatic and important enough at least to cause excited debate among the experts, yet each paper in the journals seemed to sink with scarcely a ripple of interest. The new picture of human origins raised novel and interesting questions, as well as resolving old puzzles, and it portrayed an inherently interesting subject — ourselves. Yet the palaeo-anthropologists who are supposed to answer questions about our origins seemed to be ignoring the new work. Why?

Because we worked in close proximity — actually on adjacent desks — at *New Scientist,* the baffled astronomer was able to ply a biologist with a seemingly endless stream of questions. Why did no one take the molecular evidence seriously? Where was the evidence that supported the traditional palaeontological picture? Just suppose the molecules really were telling the truth: what would the implications be? The anwers that came back were, initially, less than satisfactory: everyone with any sense knows that the fossil evidence shows the molecular studies to be wrong; the fossils prove that man and the apes diverged from their common stock at least 20 million years ago; the implications of taking the molecular evidence at face value are absurd . . . Even the biologist began to feel uneasy, and under repeated probing his answers became even less satisfactory. The more he looked for incontrovertible arguments to silence his ignorant astronomical colleague, the more he found that they did not exist; the harder he tried to fault the molecular studies, the more he convinced himself that they were founded on a solid base.

Together, we established to our own satisfaction that it is the new picture of human evolution, based on the pioneering work of Sarich and Wilson, that stands up to the closest scrutiny. It is the old picture, founded on a few petrified bones and a great deal of imagination, that turns out to be a fraud. Nobody in the palaeontological trade accepted the new picture, as far as we could tell, largely because the evidence had never been gathered together and presented as a coherent whole. Sarich and Wilson later admitted to us that they had never really gone out on the campaign trail; we decided, without asking them, to do it for them. The result, we hope, is a presentation, necessarily selective but coherent nevertheless, that should convince not only those in the trade but especially those outsiders who share our interest in human origins and who have been deprived of the best part of the story.

This book would never have come about without the peculiar juxtaposition of talents and interests made possible by *New Scientist*. For that, we must thank Bernard Dixon, the Editor who hired us, and Roger Lewin, the Deputy Editor on whose advice he acted. But we owe an even greater debt to Roger Lewin, of mingled relief and gratitude, for something he did not do. He failed to take the molecular story seriously himself, and if he had he might well have written this book – or one very like it – before we got the chance.

We are especially grateful to Vincent Sarich and Allan Wilson for sparing us so much of their time when we cornered them on many occasions over the past few months at conferences and meetings on two continents. They patiently explained not only the details of the work, but shared the incredible story of its non-acceptance over the past decade and a half. Although less than enthusiastic about the idea of a popularisation of their work in book form, they have always been open and helpful in response to our questions. We should stress, however, that the story we tell here includes our own interpretation of the evidence, and our own speculations. In no sense have Sarich or Wilson approved it, and any mistakes it contains are our own responsibility.

Other people, too, have been helpful. We are both fortunate in having spouses who are willing to take time off from their own busy lives to help us with our work; we hope they agree that the effort was worthwhile. John Reader generously allowed us to use pictures from his unique collection. And Apple Computers (with a little help from Richard Dawkins, who aside from his other talents is a programming whiz) made the whole business of writing a book like this a lot less painful than it once was.

This is also the place for a kind of apology. The English language is a marvellous tool, but it is male biased. Neither of us would ever countenance such ghastly practices as 's/he' or 'his/her', which puts us in the unenviable position of apparently ignoring half of mankind. We assure that half that it is not our intention to ignore it; that's just the way the language works. Where we say 'man', or 'mankind', or even 'his', we often mean 'person', 'humankind', and 'his/her'. Where we talk specifically about male or female human beings it should be clear that we are doing so.

Finally, we reserve our heartfelt thanks for the mass of experts in palaeo-anthropology. Their unswerving devotion to the established picture, their ridicule of the molecular evidence, and their sneering dismissals of our questions about the work of Sarich and Wilson convinced us that there must be a story worth telling.

John Gribbin
Jeremy Cherfas
August 1981

POSTSCRIPT

One evening in March 1982, scarcely a month before this book was due to be published in England, I went to hear Richard Leakey give a lecture. It wasn't so much that I thought he would have anything particularly new to say, but it is always a pleasure to listen to his excellent lectures. Besides, the event, one of the Royal Institution's elegant evening discourses in the very lecture theatre that Michael Faraday and Humphrey Davy used, was an attraction in itself. Imagine the surprise, then, of hearing Leakey tell the audience that 'the conventional wisdom,' which he had so recently espoused in his television series *The Making of Mankind*, 'was probably wrong in a number of crucial areas.' He went on to admit that he was 'staggered to believe that as little as a year ago I made the statements that I made.'

My head began to swim, but I managed to listen and take notes.

Leakey had changed his mind. He now believed that man's oldest ancestor was not the 14-million-year-old *Ramapithecus* but something much younger—indeed that '*Ramapithecus* may have been a red herring.' As long ago as 1967, when the first results of the molecular clock were published, it was clear that *Ramapithecus* could not possibly be a direct ancestor of man, for it is simply too old; but instead of accepting this information and rethinking man's evolutionary history, most palaeontologists preferred to find fault with the molecular clock. Now here was Richard Leakey, arguably the best of the young fossil-hunters, finally coming round.

It was not the molecular clock, however, that had changed his mind. It was a fresh set of fossils. David Pilbeam, the eminent Yale anthropologist and one of the clock's most entrenched opponents, had unearthed a ramapithecine in Pakistan that made it clear that these were the ancestors not of man but of the orang-utan. And other previously unrecognised fossils stored in the basements of the British Museum of Natural History and the Kenya National Museum weakened the conventional story still further.

The palaeontologists are beginning, as we said they must, to make their stories consistent with the unimpeachable evidence of the molecules. And as we also said, they are doing so not on the basis of the extensive published molecular evidence but on a rereading of the scant fossils. We could hardly have hoped for a better outcome.

Despite all else that he said during the lecture, Leakey remained silent on the crucial issue of exactly when our ancestor parted company from the chimpanzee and gorilla. He did propose that we recognize our species, *Homo,* by a new set of criteria: If it walked upright, it is *Homo;* if it had an enlarged brain, it is on the way to *sapiens.* By that yardstick the earliest marks of man are the footprints that Richard's mother, Mary, discovered at Laetoli. They are 3¾ million years old. The molecular clock dates the birth of our line at about 4½ million years old.

I managed to find Leakey after his lecture and asked him whether he now accepted the molecular clock's timings. He refused to be drawn into a direct answer. 'But,' he said, 'I think the molecular people are closer to the truth than we've ever given them credit for.'

It was an exhilarating moment.

Jeremy Cherfas
April 1982

PROLOGUE:
THE MONKEY PUZZLE

Everyone is fascinated by the mystery of the origins of mankind. Since the time of Charles Darwin, who died 100 years ago this year, it has become a well-established fact that we are descended from monkey-like ancestors, tree-dwellers who thrived in the tropical forests of 35 million years ago. But there has, until recently, been much more uncertainty about exactly how, when and where those tree-dwellers were changed, through the process of evolution by natural selection, into a variety of apes, including ourselves. This is the monkey puzzle: how did tree-dwelling monkeys become apes, when did the human line become distinct from those of the hairy apes, and what were the natural forces that caused these dramatic changes?

Evolutionists describe the relationships between species in terms of a many-branched bush, or tree, of life. Closely related species are represented by the growing tips of branches that fork from their common stem relatively recently, just below the outer surface of the bush, while more distantly related species must trace back to some remote forking in the heart of the bush to reach their common ancestor. By a happy coincidence there is a kind of tree that is popularly known as the monkey puzzle tree. Its more grandiloquent Latin name is *Araucaria araucana*, for the Araucarian tribe that bravely defended the mountain slopes of Chile where it grows. The monkey puzzle was a favourite tree of Victorians, and they called it monkey puzzle rather than Chile pine because of its leaves; described in one book as 'rather unfriendly', these are extremely sharp points that curve back along the branch, and would have posed quite a problem to any monkey wishing to climb to the tasty nuts at the branch tips. (Why the Victorians saw this as a problem we do not know, for there are no monkeys in the monkey puzzle's native forests.) More than this, one can see in the regular forking of the monkey puzzle's branches an intellectual problem for even the most agile primate, that is to trace a route out to the branches and back again. Our monkey puzzle tree — the part of the bush of life that represents the evolutionary relationships of man and his closest relatives — has proved equally difficult to trace a path through, but over the past 15 years molecular biologists have developed a set of techniques that guide us through the maze. Like Theseus unravelling Ariadne's thread to find his way through the Minotaur's labyrinth, biochemists can now trace the path taken by the molecular thread of life, DNA, which records all the twists and turns that evolution has imposed upon it. They can identify the precise relationships of the growing branch tips that represent man and the other ape lineages, and provide dates for when the branches forked.

For the first time ever, the route from monkey to man can be mapped with unambiguous precision, but the resulting chart is not the same as the one that the palaeontologists, working with a sparse and patchy fossil record, had drawn by the mid 1970s. We now know, beyond doubt, that man and the two African apes, the gorilla and the chimpanzee, shared a common ancestor only 4½ million years ago. The bush of life is 3½ thousand million years old; the three-way fork that gave rise to chimp, man and gorilla lies just below the surface of the monkey puzzle tree, 4½ million years ago. With that crucial date established, we can also identify the major changes in the environment of the world at the time of the evolutionary changes that set our ancestors on the road to humanity. And with the aid of fossil evidence we can try to establish where in the world the changes took place.

A combination of evidence from molecular biology, palaeontology and geophysics can provide the answers to all three aspects of the monkey puzzle. We know when the descendants of the monkeys set out on the road to humanity very accurately indeed. We know where the forking of the branch took place. And we know roughly why these changes took place, because we can unravel from the geophysical record the details of the environmental changes convulsing the world at that time. But the solution to the monkey puzzle made possible by this combination of evidence from across the disciplines shatters some cherished myths about the nature of humanity and our special place as Lords of Creation. Up to now it has been quite easy to accept the fact of evolution but keep man apart from it as something special. But we are not very special at all; we are all but indistinguishable from the African apes. The evidence for this dramatic new perspective on mankind, and its implications for our understanding of human evolution, are the twin themes of our book.

One

ONE PER CENT HUMAN

Man, the chimpanzee and the gorilla are so similar that it is practically impossible to tell them apart. Absurd, you say. Gorillas are hairy giants of the forest, weighing half a ton, and chimps, though smaller, are no less hairy. As for man, he walks upright, for one thing, has a large brain, and is naked. The three couldn't be easier to distinguish. But, if you were to look only at their DNA – the long molecule of heredity – you would find it a hard task indeed to separate man from the chimpanzee and the gorilla. The three species share 99 per cent of their DNA, and on that basis surely they should not look nearly as different as they do. There is only one per cent difference between ourselves and a chimp or gorilla. Yet that one per cent accounts for all of human civilisation – art, literature and science. How can such a small difference have such major implications? This is one part of the monkey puzzle. Common sense tells us that man, with his cities and societies, simply cannot really be so closely related to the African apes. Or can he?

According to the present wisdom, man has been evolving separately from the apes for at least 20 million years. That is the date favoured by most palaeontologists, and although there is some disagreement between them, with some preferring a slightly earlier date and others a later one, it is fair to say that by and large they agree that man and the apes have been biologically separate for rather a long time. Separate species means separate evolution, and the chance for DNA to accumulate separate differences. Ignoring man for the moment, tests on every other species investigated show that the longer two species have been separated, in evolutionary terms, the more differences there are in their respective DNAs. Changes in DNA occur from time to time by chance – the accident of mutation – and there is a good correlation between genetic difference, measured by comparing DNA from two species, and time. For example, in all other species that have been separated for about 25 million years, such as the dog and the raccoon, this kind of test reveals accumulated differences equivalent to about 12 per cent of their DNA. But man, the chimp and the gorilla, who according to the established story of human evolution are also supposed to have been separated for about 20 million years, differ by a mere one per cent. By comparison with other species, a genetic difference of this amount ought to correspond to a separation of man and the African apes from a common stock less than 5 million years ago.

What is going on? Are we to assume that man is in some way a special case, so that his DNA does not obey the same rules as the millions of other species that share the Earth? Or could it be that the palaeontologists might be wrong? Perhaps man and the apes have not been evolving down separate paths for 20 million years or so, but have indeed been separate for only a quarter of that time, less than 5 million years.

This is a hard idea to accept, not least for the palaeontologists whose wisdom it challenges. But it makes a lot more sense to assume that man is like all the other inhabitants of the Earth, and if necessary to rewrite palaeontological history, than to argue that man is a special case, alone in a crowd of animals.

The traditional palaeontological history is based on very slender fossil evidence. Molecular studies provide new factual evidence – the most significant development since Darwin. And, to the astonishment of most non-palaeontologists, and the embarrassment of the palaeontologists themselves, it turns out that there are no known fossils which actually contradict the molecular evidence for mankind's very recent origins. Treating man as a special case is a technique with a long tradition; but we are not interested in tradition, only in discovering the revelations that emerge from scientists' new-found ability to read genes and DNA almost as easily as you are reading this page. Ultimately, the history of evolution is written not in fossils of long-dead animals but in the genetic material of species alive today, all of which share a common descent. The differences between their genes are, literally, the differences that evolution has selected, and we would be foolish to ignore them in our search for the origins of man.

The story of man's investigation of his place in the world has been one of successive reductions in his perceived status. In the physical world, we have learned that the Earth is not the centre of the Solar System, or of our Galaxy of stars, or of the whole Universe, but is rather an insignificant planet orbiting an unspectacular star in the backwoods of a typical galaxy, one among hundreds of millions of such galaxies scattered across space. In the biological world, the process of removing man from the centre of the stage has not yet gone quite so far. After a long debate it has now become broadly accepted that man is just one animal species, subject to the same evolutionary pressures and processes as other species, but even those who accept and understand the evidence for evolution still find it hard to get away from the idea that man is somehow special, a pinnacle of evolutionary achievement.

One of the ways in which man separates himself from the rest of nature is to put his origins as far back in time as he can. If man has been evolving down his own path for 20 million years or more, he is seen as being safely distanced from the animal kingdom because he can hardly have been on the road to humanity for so long without leaving behind almost all his animal heritage. Perhaps that is why each new discovery of man's oldest ancestor is greeted with such enthusiasm; once the idea of an ancestry had been accepted at all it rapidly became the case that a longer ancestry was more acceptable than a shorter one. Some people still cannot accept the evidence that we are products of evolution in the same way that other animals are; even people who do accept the evidence of evolution seem instinctively to wish to keep our relationship with other animals as distant as possible. The molecular evidence disturbs this view by shortening our ancestry dramatically, and that is surely one reason why palaeoanthropologists are so unwilling to accept it. Sherwood Washburn, professor of physical anthropology at the University of California at Berkeley and an inspiration to many of the new breed of molecular anthropologists, once said, 'I suspect that if molecular anthropology had shown man and apes to be very far apart, the concept of a correlation between genetic difference and time would have been accepted without debate.'[1] What sticks in the throat even of the palaeontologists is that we are not even distant cousins to hairy apes but brothers − next of kin to the chimp and the gorilla.

For the ordinary person, with no scientific training, the anthropocentric viewpoint that Washburn decries is understandable. Each of us is bound to regard himself as the most important product of the evolutionary process, and from that point of view it is only a small step to thinking of mankind as the species at the top of the evolutionary ladder, the ultimate goal of all the evolutionary changes that have taken place in life on Earth. But this viewpoint is as distorted as the idea of a flat Earth, which also seems so logical from our everyday experience. Evolution does not have ultimate goals – except for the replication of DNA. In terms of success every species now alive on Earth is successful, while every species that went extinct was unsuccessful. Man is no more special in the biological world than our Sun is among the stars. We are different from other species, certainly, in being aware of these puzzles and able to read and write books about them. But there is no absolute sense in which our particular skills of intelligence and adaptability can be regarded as superior to the special skills which enable ants, say, or bacteria to survive and reproduce in large numbers on Earth.

So who are people? We believe that even today an unconscious thread of anthropocentrism runs through a great deal of discussion, teaching and writing on human evolution. It is very difficult to try and forget that we are human, and to assume the role of an extraterrestrial scientist dispassionately studying the evolution of life forms, including the human life form, on planet Earth. Charles Darwin appreciated the problem, as the quote on page *v*, taken from the beginning of Chapter One of *The Descent of Man*, shows. We have followed Darwin's advice, treating man as just one species among many, and in doing so we believe that we have come up with a picture of ourselves that has as much capacity to surprise in the 1980s as Darwin's ideas had in the 1860s.

Man is quite obviously related to the great apes, a fact that was clear not only to Charles Darwin and other early naturalists but also to the people who live among the apes; in South-East Asia 'orang-utan' means 'man of the forest', and many of the African tribes that know chimpanzees and gorillas refer to them as sorts of men. Some of the features that we share with chimps, gorillas and orangs are clear, as are some of the differences. But there is one cluster of characteristics that is not so obvious. If you had to tell a visitor from another planet what distinguished man from the apes you might mention man's habit of standing upright and walking on two legs, and you would probably also describe man as having a larger brain. Both of these are true, though the unique shape of man's head is as much a matter of his small, squashed face as of his large brain. But between the overblown brainbox and the re-modelled pelvis, the similarities are much more overwhelming than the differences. From neck to hips we share with the apes a specialised anatomy that unambiguously identifies our simian ancestry. These similarities, so obvious once they have been pointed out, are another aspect of the monkey puzzle that we will be trying to account for.

Picture an orang-utan or a chimpanzee climbing through the trees; now imagine a human gymnast on the parallel bars, the rings, or the horizontal bar. Think of a chimpanzee brandishing a branch as it mock charges an intruder, and a human about to throw a javelin. Think, quite simply, of getting up in the morning and stretching. All these activities, and plenty of others, reflect the very special torso of the apes, a torso shaped by natural selection to produce an animal that can hang by its arms from branches and stretch sideways. It is the upper body that sets the apes apart from the monkeys, and it is the upper body that makes the apes seem so 'human'. The things apes do that remind us of ourselves, and keep us riveted at the zoo, are almost all a product of the torso that we share with them. Monkeys, too, are fascinating, but the reason they don't seem nearly as human as apes is that they don't have the apes' upper body.

The specialised anatomy of the apes was probably selected because it allowed the evolving animals to get at fruits growing right out at the tips of branches. A light, four-legged monkey is an astonishing athlete, and can run with ease along the tops of slender branches, but it cannot get right out to the tips of those branches, and that is where the fruit is most abundant. An animal that can hang beneath a whippy branch is in a much more secure position than one balancing above the branch, and if the animal can also stretch sideways it is even better off, because it can get at the fruit without having to rely on the thinner wood near the fruit. That, in turn, means that the animal that can hang around can be larger, and in general there are many advantages to being bigger.

The anatomical solutions to the problem of hanging from branches are many and complex, but we can mention a few. By contrast to monkeys, the collarbone of apes (and, of course, man) is longer and helps to keep the shoulder away from the chest. The shoulder itself is much freer and can move in all sorts of ways: it isn't much of an effort for you to scratch your right ear with your left hand from behind your head, but no monkey could do this. The arm can move freely within the shoulder joint: you can put your palm flat on a table and rotate your arm through 180 degrees without moving shoulder or wrist. Your hand can bend through a full right angle in the direction of your little finger, and in common with the apes you have strong biceps muscles that enable you to pull yourself up on a bar. All these anatomical adaptations, shared by man and the apes, fit perfectly with a mode of travel that depends on swinging along under branches. The pivot of the hand about the wrist allows the body to swing forward in the direction of travel while still securely held. The strong muscles of the upper arm and trunk help to power the swinging, and the enormous freedom of the shoulder joint allows manoeuvrability even while hanging below the branch.

This cluster of adaptations that allows movement by arm-swinging, or brachiation as it is called, is a property only of the apes — gibbons, orangs, gorillas, chimpanzees and men. But the role of brachiation in the evolution of the apes has been much misunder-stood. It is quite clear that the apes don't necessarily spend all, or even most, of their time in this sort of movement, which led many people to argue that adaptations for locomotion are not important in the evolutionary history of man and the apes. The essential point is not that the apes all *do* brachiate, but that they all *can* do so when the need arises. No monkey can, or does. An experiment by Virginia Avis illustrates beautifully the point we are making.[2] Sherwood Washburn persuaded Avis that it would be a good idea for her to test his theory that brachiation was an important behaviour separating monkeys from apes. Funded by the Ford Foundation, and with the help of the Chicago Zoological Society and the workers at Brook-field Zoo in Illinois, Avis built a special cage in which to observe her subjects. The cage was equipped with a variety of supports that would allow the animals a great deal of freedom of choice in how to move around the cage. There were thin bars of bamboo and wild grapevine that were flexible and wobbly, and there were other bars, the same diameter but made of rigid metal. Wide rough branches were also present. The cage therefore had branches small enough for monkeys to grasp, and some of these were bendy while others were rigid. The subjects were three species of ape — the gibbon, orang-utan and chimpanzee — and 12 species of monkey, and Avis watched these at play in her well-equipped primate gymnasium for long hours. She also watched them, and a large number of other species, including the gorilla, in their home cages to ensure that her data were truly representative. In all cases she saw that the monkeys were very unwilling to use the flexible supports; even when they were very frightened the monkeys would go the long way round rather than walk along the swaying bamboo, and a hungry monkey would ignore food if the only way to get at it was along one of the flexible supports. The apes, by contrast, were perfectly at home even on the flimsy vine, which they crossed with ease by dangling below it. The monkeys inevitably ran on all fours above the support, and the apes almost always swung below it. If, however, the branch was large and stable then the apes too stayed on top, the gorilla, chimp and orang walking on four limbs and the gibbon on two. Apes brachiate, monkeys do not.

Man, of course, fits the ape pattern rather well, as anyone watching children on the ill-named monkey bars would readily agree. It is possible, but not easy, to say that these similarities are a coincidence, the result of the process called parallel evolution. As Avis herself said, summarising the implications of her important results, 'to contemplate deriving man from a hypothetical "brachiating" monkey is to forget the vast number of features, structural, embryological, physiological and psychological, which man shares with the ape . . . For this reason one may conclude that if in the future a brachiating ancestry for man becomes a proven fact, then there is little doubt that he shared it — for perhaps millions of years — with his fellow apes.'[3]

So the main anatomical similarity binding the apes and man together and setting them apart from the other primates is that they are brachiators. How we came to share those features is in a sense the question of our book. Many anthropologists have been content to turn a blind eye on this question, saying simply that man and the apes evolved their distinctive similarities more or less by coincidence as they swung from branch to branch through their separate evolutionary trees. Washburn, however, has never had time for this explanation. 'There is a profound similarity in the motions of the arms of man and apes,' he points out, 'and on any playground one can see humans brachiating from bars, hanging from one hand, and exhibiting a variety of motions and postures which are similar to those of the apes.' Recalling the experiments of Avis, his student, Washburn says categorically, 'man is still a brachiator. He is simply the one who is least frequently in the situation which calls forth this behavior. Our legs are too heavy, and our arms are too weak for efficient brachiation but, when we climb, we climb like apes and not like monkeys.'[4] Washburn not only sets out the puzzle, he also comes up with a simple answer: we and the apes share a common ancestor who was a brachiator, and because we are all descended from that brachiating ancestor we are all brachiators too. No one would disagree with this, were it not for the fact that there are no good brachiators among the fossils of 20 million years ago. Washburn's answer to this is even simpler. There are no fossils because brachiation itself did not evolve until about 15 million years ago, and man and the apes separated very recently indeed. Obviously the two stories cannot both be true.

This, perhaps, is the place to clear up a common misconception. Nobody suggests that we are descended *from* the hairy apes! What evolutionary studies reveal is that man and the apes *share* a common ancestor, a brachiator perhaps superficially more ape-like in appearance than man-like, but strictly speaking neither modern man nor modern ape. You and your sister share common ancestors — your parents — but that doesn't mean that you are descended from your sister, nor she from you. Normally when there is some uncertainty over the evolutionary past of a group of animals, the scientists turn to the fossil remnants of that past to sort out the history. The problem, for man's own history, is that although there are quite a few fossil primates the ones that would settle the matter don't seem to have been found yet. There is, quite simply, no good fossil evidence of any brachiating ancestral ape. Now it is true that there are some quite notable gaps in the fossil record, and to overcome the embarrassing lack of evidence many palaeontologists have consigned the interesting bits of the story to those gaps. For example, there is a dearth of fossilised primate remains from between 15 and 22 million years ago, and so experts such as Richard Leakey, son of Louis and Mary Leakey and a fine palaeontologist in his own right, would have us believe that it was just then that the line leading eventually to man separated from the line leading to apes. This may be convenient, in that no evidence is around to spoil the hypothesis, but it raises more questions than it answers. For one thing, the fossil apes from this period, the early Miocene, certainly do not show clear evidence of being brachiators, and yet their descendants, man and the apes, are indubitably brachiators. Are we to believe that a non-brachiating primate gave rise to man and the apes which then evolved along separate but parallel paths for 20 million years to end up with such similar anatomies? This is what Leakey asks us to believe, but it just isn't reasonable. Nor is it supported by the evidence. Far better to suppose that a properly brachiating ape, descended from a Miocene forebear, was the common ancestor who gave rise both to the apes and to man much more recently.

This, we believe, is the correct story: A successful Miocene brachiator gave rise to the gibbons 10 million years ago, the orangutan 7 million years ago, and finally split three ways — into man, the chimp and the gorilla — 4½ million years ago. And unlike all the other theories of man's origins, this one has the advantage of a great deal of easily understood and interpreted evidence. The key to this new view of mankind, its origins, and history, is the development of refined techniques to probe the nature of life at the molecular level. In terms of human evolution, the molecular evidence reveals who we are. It tells us just who our closest relatives among the primates are, and just how closely related to them we are. It also tell us exactly how long it is since the human line and the lines which lead to our closest relations diverged.

It is all very well to talk of refined techniques and new information, but what are these new techniques, and how do they produce new information? The details will come in succeeding chapters, but in order to set out our case we also have to set out some preliminary explanations. The essence of the approach is simplicity itself, and we have already summarised it. Evolution proceeds by the modification of existing forms of life. The modifications are carried in the form of changes to the DNA, which makes up a chemical storage library of the information needed to build a particular member of a particular species. One definition of a species is that its members can breed with one another, which means, effectively, that they can exchange bits of DNA. When two individuals are in different species they can no longer share DNA. This means that any changes that accumulate in the DNA of one species can spread to all members of the species, but not into other species. When one species splits into two new species the offspring at first share the DNA of the parent species, but as each species evolves it accumulates new changes in its DNA that are unique to itself, and the two species diverge from one another. So if we could look at the differences between the DNAs from two different organisms, whether the two were from the same breeding population of a single species or from widely separated species, we should be able to count up the differences and use them as a measure of how closely related the two individuals are. The more similar their DNA, the more closely the two are related. More interesting, and for our purposes more useful, the more similar the DNA the more recently they shared a common ancestor.

It sounds simple enough in theory, and today, with the technological advances that have been wrought in the field of genetic engineering, it is pretty simple in practice, but the idea is not especially new. One of the first people to demonstrate that molecules hold the key to the past of the animals carrying them was George Henry Falkiner Nuttall, professor of biology at Cambridge University. Nuttall was born in San Francisco in 1862, and was educated throughout America and Europe. He did his doctoral work in Germany, where he came into contact with the revolutionary ideas of Paul Ehrlich. Ehrlich, who all but invented the science of immunology and pioneered the use of antibiotic 'magic bullets' against diseases, noted that blood contained a substance − precipitin − that would coagulate certain proteins. And if small amounts of foreign proteins were injected into an animal then the animal would manufacture the precipitating antibodies, as they were called, and from then on could cope with a second invasion of the foreign protein, or antigen. This discovery forms the basis of immunisation against disease. Making antibodies takes time, so the doctor injects a harmless preparation of the organism responsible for the disease. The patient's immune system responds to the disease antigens by producing the appropriate antibodies. Then, if the patient is later invaded by the harmful disease organism, the antibodies attach to the invaders and guide other cells to destroy them before the invaders have had a chance to do much damage.

Ehrlich discovered that antibodies were specific to their antigen. An antibody that protected guinea-pigs against one toxin, ricin, would not protect them against abrin, a different toxin. This same specificity is the reason why immunity to, for example, measles is no protection against chicken pox. To explain this specificity, Ehrlich borrowed the idea of a lock and key from Emil Fischer, a colleague working on 'specific ferments' − what we would now call enzymes. Ehrlich said that the antibody and antigen had complimentary shapes, so that they fitted together like a lock and key: antibodies − locks − manufactured in response to one antigen would not work with keys − antigens − of a different shape. He further speculated that the antibodies made by one animal when injected with the blood proteins of another animal would react only with blood proteins from animals of the same species as the original donor species. Another German, Paul Uhlenhuth, did some work along these lines and gave rise to the science of forensic immunology, but it was Nuttall who really took up Ehrlich's speculation with a vengeance.

His justification was straightforward. 'The persistence of the chemical blood relationships between the various groups of animals serves to carry us back into geological times, and I believe . . . that it will lead to valuable results in the study of various problems of evolution.'[5] Nuttall knew nothing about DNA, of course, for it was still more than 40 years from being finally implicated in the mysterious business of inheritance, but that mattered not at all. We now realise that the structure of proteins depends absolutely on the structure of DNA, so that the proteins of two animals differ only because their DNA differs to the same degree; but for Nuttall all that mattered was that he could get a variety of immune reactions to his various bloods and antibodies.

In just a couple of years Nuttall conducted 16,000 experiments on no fewer than 900 blood specimens from a vast number of species. Most of the blood samples were supplied through the 'generous aid of some seventy gentlemen around the world.'[6] These included the Hon. N. Charles Rothschild, who supplied 89 samples of animal bloods and enlisted many of his huntin' and shootin' friends to Nuttall's cause. Nuttall provided his collectors with materials and instructions, advising them to soak blood or serum onto a strip of filter paper about 5 inches by 3 inches, which could then be dried by pinning 'against the edge of a table or shelf or upon the branch of a tree. Where it was impossible to wait for the strips to dry, as when out shooting,' Nuttall tells us, 'collectors were requested to place each sample separately in the paraffined paper covers which I supplied.'[7]

The 16,000 experiments were perhaps not so laborious as they sound, for each consisted only of dissolving the blood from a portion of filter paper and then adding a drop of one of the many antisera that Nuttall had prepared from various species, from fishes to man. Rack upon rack of tiny test tubes could be prepared in this way, and allowed to stand for 24 hours. During this time Nuttall examined each tube closely to see whether there was any evidence of a precipitation reaction. The results − reactions only with bloods of closely related species − were generally in full accord with evolutionary trees based on fossils and comparative anatomy, which vindicated Ehrlich and Uhlenhuth, and contributed in no small degree to Nuttall's election to the Royal Society in the year his book was published.

Naturally Nuttall also investigated his own species − indeed this was one of the first of his tasks – and in a lecture he gave to the London School of Tropical Medicine on 28 November 1901 he summarised his investigations of the primates thus: 'If we accept the degree of blood reaction as an index of blood relationship within the Anthropoidea, then we find that the Old World apes are more closely allied to man than are the New World apes, and this is exactly in accordance with the opinion expressed by Darwin.'[8]

His 16,000 experiments had convinced Nuttall of the value and validity of the technique, but he recognised that it was only a beginning. 'I hope that the work done will stimulate many to further investigate the many problems which present themselves,' he wrote.[9] 'My studies must be regarded in the light of a preliminary investigation, which will have to be continued along special lines by many workers in the future.'[10]

It is astonishing, from the perspective of the 1980s, when the extension of his technique is seen as so revolutionary that palaeontologists have yet to come fully to terms with it, that Nuttall did his work barely 40 years on from the publication of *The Origin of Species*. Unfortunately, despite the obvious promise of his approach, Nuttall's hopes were not realised for quite some time. There were occasional sporadic attempts to carry his work further, but it was not until the late 1950s that the torch that Nuttall kindled was once again fanned to life. More than anyone else, the man responsible was Morris Goodman, now professor of anatomy at Wayne State University School of Medicine in Detroit. Goodman, a greying, well-groomed and slightly portly man who would seem to be more at home in a bank than at a laboratory bench, used a refined version of Nuttall's precipitin reaction to investigate the affinities of a number of primate species. Where Nuttall had simply mixed antiserum with antigen and estimated the cloudiness of the reaction, Goodman used the much more sensitive process of immunodiffusion, developed by Swedish scientist Örjan Ouchterlony. First Goodman made plates of agar, the jelly-like substance extracted from sea-weeds and used by bacteriologists as a medium on which to grow their cultures. Then, into the plate of agar he cut three wells so as to leave a triangular block of agar in the middle of the plate, bounded on each side by a well. Into one well he put the antiserum, and into the other two he put the antigens he wished to compare. The antibody and antigen molecules diffuse through the agar block in much the same way that a wine stain spreads on a white tablecloth, and eventually come into contact with one another. Where they meet the precipitin reaction occurs and a white line forms in the agar. If the two antigens are identical in shape and size they will travel through the agar at identical rates and the lines that each forms will join in the middle of the triangle. If they are dissimilar then the lines do not join completely, and it is possible to tell how dissimilar the antigens are by the pattern of precipitin lines. Comparing a number of species in this way, Goodman discovered that the antigens of man and chimp are practically identical, while those of the gibbon are a little different. Next in the affinity to man are the Old World monkeys, then the New World monkeys, then the lorises and then the lemurs. As Goodman said, 'this sequence fits in well . . . with the fossil record'.[11] But so far he was talking only about the order of the relationships — no specific dates had come into the story yet.

It was in an addendum to his 1962 paper that Goodman gave the first hint of the bombshell he was about to drop. He had obtained serum from the orang-utan and gorilla to add to his extensive samples, and had compared them with man. Although the work was by no means complete he chose to record his preliminary observations. 'Man is closer to the African apes (gorilla and chimpanzee) than to the Asiatic apes (gibbon and orang-utan),' he wrote, which was perhaps not such a surprise as everyone was by now agreed that man emerged in Africa. But, he continued, 'gorilla and man appeared to be almost identical, chimpanzee diverged slightly from man, and gibbon and orang-utan each clearly diverged from man.'[12] Other proteins showed chimp closer to man, and on a number of subsequent tests it proved impossible to put either chimp or gorilla unequivocally closest to man. 'There is no indication from the serological data of any closer relationship between chimpanzee and gorilla than between either of these apes and man' was how Goodman put it when he'd finished his extensive series of comparisons.[13]

The evidence established that the African apes and man are mutually closely related, each one equally close to the other two, and by contrast to the earlier data this was something of a shock. At the time, as now, everyone believed that the chimpanzee (*Pan troglodytes*) and gorilla were much closer to each other than either was to man. This is reflected in their biological classification: with the orang-utan they form a separate family, the Pongidae, while man alone occupies the family Hominidae. Goodman has revealed the falseness of this classification. 'Evidence is provided that the phyletic

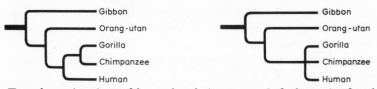

Two alternative views of the apes' evolutionary tree. Left, the version found in most modern textbooks, in which the human line is separate from the common ancestor of gorilla and chimpanzee. Right, the molecular evidence places man, chimp and gorilla together. There is no human ancestor that is not also an ancestor of chimp and gorilla. Note that these trees give no indication of timescale, simply the order of branching. The molecular clock is calibrated in other ways (see page 127).

line which branched to give rise to *Gorilla, Homo,* and *Pan* did so after it had separated from more ancient lines leading to *Hylobates* and *Pongo*,'[14] said Goodman in the dry unemotional language of a scientific paper, effectively countering the accumulated wisdom of the ages — that is, of the century since Darwin — there and then. Beliefs, however, die hard, and although Goodman has done much to alter our perception of man's evolutionary past he has not yet succeeded in toppling the classification and replacing it with what the DNA demands, no less than a new subfamily, the Homininae, which would include only the African triumvirate: man, chimpanzee and gorilla.

Goodman has very effectively redrawn the route of man's path from an ape to his present position, and astounded several people with the details of the forks along that route. A journey, however, is much more interesting if one knows not only where one is going, but when one will get there. Goodman's data tell us only where man has been. It was left to others to discover the times at which he passed each signpost.

In the early 1960s, as Goodman was getting under way with his work, Vincent Sarich was just starting his postgraduate research as an anthropologist at Berkeley. One of Sarich's teachers was Sherwood Washburn, who naturally acquainted Sarich with the puzzles that surround man's anatomy. Both men felt that Goodman's work might provide an answer if it could be shown that man was not only closely related to the great apes but also had not long been separated from them. 'In those days the field was small,' Sarich told us, 'and you could read it all up in a couple of weeks, so I agreed to prepare a seminar on the molecular evidence for one of Washburn's classes.'[15]

He did so, and with his chemical training realised that there was far more that could be done with the molecules. Washburn sent him off to see Allan Wilson, a biochemist just setting up at Berkeley; together, Sarich and Wilson became one of the major forces of molecular anthropology.

Sarich went to work in Wilson's lab, and soon had results that set them thinking. Using techniques vastly more sophisticated than those of Goodman, he carefully measured the similarities of one of the proteins commonly found in blood, albumin. So precise were his measurements that Sarich could detect a difference of a single amino acid in several hundred, which enabled him to give each of his species an immunological distance from any other species. Like Goodman before him, Sarich found that the albumins of man, the chimpanzee and the gorilla formed a unit distinct from those of the Asiatic apes. He went on to use his immunological data as a clock, measuring the time that had passed since man and the other African apes had parted company. The method is simple: the degree of difference between the various albumins is related to the time since the species split, so if one has a fairly accurate date for one split, one can use the albumin differences to date the other splits. In their 1967 paper in the prestigeous journal *Science,* Sarich and Wilson took as their reference point the split between the Old World monkeys and the apes, which the fossil evidence suggests 'split about 30 million years ago'. With this benchmark, 'the time of divergence of man from the African apes is . . . 5 million years'.[16]

Not 20 million years ago in the Miocene but only 5 million years since we and the apes shared a common ancestor. Only 5 million years since our DNA and theirs resided in the same cells. It was an audacious suggestion, bound to stir controversy, but it solved the monkey puzzle. 'If the view that man and the African apes share a Pliocene ancestor . . . is correct,' Sarich and Wilson wrote, 'a number of the problems that have troubled students of this group are resolved. The many features of morphology, particularly in the thorax and upper limbs, which man and the living apes share in varying degrees, but which were not present in the Miocene apes . . . are then seen as due to recent common ancestry and not, as generally accepted, to parallel or convergent evolution.'[17]

In essence, that's all there was to it. A molecular clock that, when set ticking against the evolution of the human species, revealed our very recent emergence as an independent line. It makes a lot of sense, and since those first publications the evidence has continued to accumulate. Moving on from immune reactions to proteins to the DNA itself, which orders the shape of the proteins, Wilson and another student, Marie-Claire King, showed that chimp and man are 99 per cent identical.[18] This, and an even greater collection of evidence, makes it extremely unlikely that man and chimp have been separate for anything like 20 million years. In fact, the accumulated evidence now points to a split only 4½ million years ago.

In a perfect world, one in which scientists actually behaved as the automata they are popularly caricatured to be, that would have been the end of the story. All biologists everywhere would have welcomed the new evidence and would have set to with a will to see whether they could now make more sense of the fossils. Unfortunately, the world is far from perfect, and the evidence of the molecular clock, at least as far as it applies to man, has not been welcomed as it might have been. Just why this is so may become clear by the end of this book, but we do not wish to berate palaeontologists and others for not accepting the data before them. Instead, we will provide the evidence and hope that science, rather than sociology, prevails in the end, for it is undeniable that people are African apes, and that but 5 million years ago only one of the apes in Africa was destined to leave any survivors. This is the dramatic revelation of the molecules.

The drama of these discoveries lies in the drastic way they change the picture derived from the fossil record. Until very recently, as we've said, biologists were confident that the split between man and his closest relatives occured some time more than 20 million years ago, and that chimp and gorilla were more closely related to each other than either was to man. Now, the molecular clock tells us that the split occurred less than 5 million years ago. And although we knew that the African apes were our closest relatives, the molecular evidence tells us that they are much more like us, in terms of the DNA that ultimately runs their bodies and ours, than anyone had suspected. Remember, the blueprints from which one might construct a chimpanzee or a gorilla differ by less than one per cent from the blueprints for a human being.

We are, more clearly than ever before, established as members of the ape family. The description of ourselves as brothers – and sisters – to the apes is indeed apt, for we are as closely related to them as it is possible to be without being members of the same biological species.

How Darwin would have welcomed this evidence. How it would have confounded his critics.

But just as Darwin had to fight against the scientific establishment of his day to have his ideas accepted, so the traditionalists today, brought up on fossil evidence and the old picture of evolution, are reluctant to accept the new evidence. They prefer the fossils, even though these appear to clash directly with the molecules. The story is familiar throughout the history of science. Yesterday's revolutionary new concept becomes today's dogma, and each generation is as reluctant to accept new ideas as the last. The establishment never likes to rewrite the textbooks. But, clearly, something has to go. The fossil evidence is much weaker than the molecular evidence. If only one version of history can be correct, there is no doubt in our minds which one it has to be. And that makes the whole story of human evolution very different from everything that has been taught so far.

Who are people? People are apes; we are, genetically speaking, as closely related to our brothers the chimp and gorilla as they are to each other, and the special features that make us human only *began* to emerge onto the evolutionary stage 4½ million years ago. The traditional timescale of human evolution – the timescale enshrined in textbooks and popularised both on TV and in a flood of books marking the centenary of Darwin's death – is, quite simply, wrong.

But, before we explain in detail the revolution which is now transforming the study of human origins, it is only fair to describe how the foundations of evolutionary understanding were provided by the fossil evidence. With all its flaws, and they are many, the traditional picture of human evolution remains a remarkable achievement of human intellectual endeavour, constructing the outlines of a theory of our origins from a handful of bones.

The new picture supersedes the old, but it does not conflict with the fossil evidence. Indeed, the fossil evidence supports the new molecular picture of human origins rather better than it supports the traditional picture, as examples like Richard Leakey's consignment of the man–ape split to the fossil gap 16 million years ago emphasise.

Two
THE EVOLUTIONARY CLOCKWORK

It is still traditional for descriptions of human origins and evolution to start out from the fossil evidence that our ancestors were different from us, and to use this as a jumping-off point to attempt to explain how and why we evolved. But this approach makes the mistake of many attempts to explain science, by taking the reader laboriously through the history of science and covering not just the successes of our scientific predecessors but also — and often in confusing detail — their mistakes. Today we have an enormous advantage over Charles Darwin, not only in that we have a comparative wealth of fossil remains to talk about but, far more important, we have a good idea of what makes evolution tick, at a molecular level. The story of life on Earth is the story of the life molecule, DNA, and the many disguises it wears, in the form of living species, in order to survive. Because we have the benefit of hindsight we can leave aside the blind alleys of scientific investigation and stick to the main highway of progress. That is nothing to be ashamed of, and we see no reason to pretend that we follow the often painful reasoning process which lead to the modern scientific understanding of evolution before we pull the best theories out of our metaphorical hat to astonish you with our cleverness. We prefer to *start* with the best modern picture of the workings of the evolutionary clockwork before moving up from the molecular level to the scale of whole species to see how changes to the DNA have wrought changes to the organisms in which the DNA resides. It is, after all, easiest to build a house from the foundations upward, and DNA is the foundation of evolution.

This does not mean that we should ignore the existence of species altogether, even at this stage. The sheer diversity of life on Earth is staggering, with millions of species discovered, described, and classified, and the likelihood of millions more still unclassified late in the 20th century. This diversity is, in one sense, just a shadow of its former glory, because at intervals throughout the history of life great catastrophes have wiped out huge numbers of species. At the end of the Permian era, some 225 million years ago, something happened that killed off 96 per cent of all species alive at the time, and the

Permian catastrophe was just one of many great dyings. The diversity we see around us has evolved from the 4 per cent of species that made it through the Permian catastrophe, and the equally small numbers that somehow survived all the subsequent catastrophes too: the species living today probably represent less than one tenth of one per cent of all the species that have ever lived on Earth. To this seeming chaos, the theory of evolution by natural selection, arrived at independently by Charles Darwin and Alfred Russel Wallace, brings order. And beneath the order of evolution lies an even deeper order, that of the molecular code of life, as it is spelled out in DNA.

Even without any knowledge of the variety of life that once existed on the Earth, the variety that they could see around them every day was too much for philosophers to explain scientifically throughout most of human history. It is hardly surprising that early philosophers found comfort in the notion of a special creation, in which some benevolent force populated the Earth with all the creatures that exist and set man in a special place above them. Today, many people get similar comfort from the ideas of authors such as Erich von Däniken, who argue that the Earth was seeded by visiting spacemen, even though these authors don't say who seeded the spacemen's home planet. Many more take comfort in an extremely literal interpretation of just one creation myth, the one described in the Christian Bible. All societies have myths of creation – we all want to know where we came from, and why. But the Christians' special creation is not a scientific theory, it makes no predictions and can never be tested or disproved, the twin hallmarks of a genuine scientific theory.

By definition, an act of faith has no basis in logic, and no arguments of ours can be expected to overturn anyone's faith in the Biblical myth of creation. Nor, in a way, would we want to, if belief were all that was at stake. Subscribers to any religion should, of course, be free to take comfort in the creation myth of their choice; by the same token those of us who take comfort in scientific logic and the inescapable beauty of well-tested theories of human origins should be equally free from censure and censorship. Creationists often accost us with the claim that the theory of evolution is 'only a theory' and that therefore 'it doesn't even pretend to be the true story', but this misconception is based on the everyday use of the word theory, not on its use in a scientific sense.

To a scientist, an idea that is only a speculation about how the

(33)

natural world might work is called an hypothesis; the idea is only graced with the name theory *after* it has been proved to be a good description of the natural world. That does not mean it cannot be superseded — Newton's theory of gravity has been improved upon by Einstein's theory — but future modifications are likely to be refinements rather than an overturning of the whole applecart. For most of us, Newton's theory is an entirely adequate theory of gravity, and will serve, for example, to get men to the moon or a space probe to Saturn; Einstein's refinements only become important if you want to explain, for example, how light is bent as it passes near a star. Future refinements of Darwin's theory of natural selection are likely to be comparably esoteric; as far as the theory of natural selection is concerned the first of five definitions of the word theory to be found in the *Shorter Oxford Dictionary* is surely the relevant one: 'a hypothesis that has been confirmed or established by observation or experiment and is propounded or accepted as accounting for the known facts'. Fundamentalists use the word in its everyday sense, the fifth meaning listed by the OED, applying it, as the dictionary says, in a 'loose or general sense' to mean 'a mere hypothesis, speculation, conjecture'. Darwin waited more than 20 years before he published his theory of evolution, which indeed started out in his mind as a speculation or conjecture but which had grown into a theory, in the scientific sense, as a result of those years of painstaking study during which he assembled the facts to support his case. In essence, the theory is simple.

There are just three basic ideas behind the theory of evolution. One is that animals and plants differ from one another and that these differences can be transmitted to their offspring. This is not simply the idea of 'like begets like', which tells us that a bitch will always give birth to puppies and never kittens. Rather, it recognises that a dog with short legs is more likely to have offspring that also have short legs. The second component of evolutionary theory is the idea of a struggle for survival, which both Darwin and Wallace developed in terms of a competition for limited resources, each of them doing so after reading Thomas Malthus' essay 'On Population'. Malthus pointed out that populations always grow to the limit of available resources, because the number of individuals able to breed keeps increasing until the population is cut back. All animals have a potential for reproduction far greater than that appropriate to maintain the stable level at which the population is in balance with food supply, or available space, or whatever resource is in short supply. The result is that many individuals must die in each generation. The third postulate of natural selection is that those members of the population which are — for whatever reason — slightly better equipped than the others to make use of the scarce resources will produce more offspring. Members which are less well equipped will produce fewer offspring and over many generations the better-equipped come to dominate, establishing the normal characteristics of the species. If a variation arises — an animal with sharper teeth, perhaps — and the variation is more succesful at obtaining food, say, then over many generations what was once a variation becomes typical of the species. New variations arise all the time, and conditions — such as climate — change so that different adaptations may be successful, so the whole process is endless. Darwin and Wallace argued that the cycle of natural selection was all that was needed, operating over the very long history of life on Earth, to produce all the species of life from one common ancestor. The reason they shocked the Victorians, however, was not that they suggested that life in general was involved in an evolutionary process, but that they insisted that mankind was no more than one species among many, just another product of evolution by natural selection. What neither Darwin nor Wallace knew was that the mechanism by which variations on the evolutionary theme occur is that of changes in the DNA blueprints that describe the forms of species; and neither Darwin nor his opponents in the Victorian Church knew that the DNA blueprints describing a human being differ by just one per cent from the DNA blueprints that describe a chimpanzee or a gorilla.

So familiar is the name DNA today that we were content to launch straight into our story of the dramatic new discoveries of mankind's close kinship with the African apes confident that although the details of genetics and inheritance might not be familiar the concept of DNA — deoxyribonucleic acid — would not be unknown. You don't have to know much about evolution to appreciate the drama of the discovery that 99 per cent of the DNA in your body is the same as the DNA in a chimp's body. But now we have caught your attention, and you want proof; to provide that proof we have to go into a little more detail about the structure of DNA itself, and its role as a component in the clockwork of evolution, ticking off the variations on the evolutionary theme.

DNA is a long molecule built up as a chain of many smaller subunits joined together — a biopolymer. If all the DNA in a human cell were stretched out and laid from end to end it would be about 175 centimetres long but less than a millionth of a centimetre thick; to put it another way, if we increase the thickness to that of an average violin string, the molecule would be almost nine miles long.[1] Inside the cell the long molecules are coiled and twisted into tiny compact strands. The links of the chain are made of chemical units, groups of atoms which are tightly bound together and act as the individual building blocks of the DNA. Alternate groups along the chain are formed by a phosphate unit and a sugar called deoxyribose, which gives the molecule its name. And each sugar group has another type of chemical unit, called a base, attached to it and sticking out sideways from the long spine of the DNA molecule. The bases are the key to DNA's role as carrier of the genetic code of life, for they come in four varieties that between them make up the alphabet of the code.

There are two sorts of base, called purines and pyrimidines, and there are two types of each. The purines are called adenine and guanine, and the pyrimidines are thymine and cytosine. For simplicity they are usually referred to by their initials — A, G, T and C — and these four letters are the alphabet of DNA. Along a single DNA molecule 3.8 cm long (the average length of one human DNA molecule) there will be some 10 million sugar units, each with its base attached, and the whole string makes up an elaborate coded message which may run . . . ATA ATT CGT TCA AGG GCT ATA TGC CGT TTA AAA CGT . . . almost endlessly along the molecule. A four-letter alphabet is, of course, ample to write down all the information needed to build a human being, provided the message is long enough. It may seem modest compared with the 26 letters of the English alphabet, but the way the message is passed on is exactly the same as the way our own message is being passed on to you, in the form of a long string of letters which just happens to be printed on paper rather than strung along the links of a molecule of DNA.

The building blocks of the genetic material are the same in all forms of life on Earth, with the exception of some viruses that use a different sugar. This hints strongly that all life on Earth has evolved from some original living DNA that appeared early in the history of the Earth — but that story lies outside the scope of the present book. The DNA itself, however, the arrangement of those building blocks, differs from one species to another and, to a lesser extent, from one individual to another within the same species. These differences are the fundamental facts on which our book is based. All individuals carry coded DNA messages unique to themselves which describe the construction, operation, care, and maintenance of their own body, and obviously the plans for one animal, a man say, are more similar to those of another man than to the plans for a completely different species. But the details do differ, even between siblings (except of course for identical twins, produced from the splitting of one fertilised egg and as a result sharing identical copies of the DNA message).

Just as the string of letters making up our message to you is broken up into words and separated by punctuation marks, so the message of life coded on the strands of DNA is broken up into words and separated by punctuation marks. How does the code operate? The workhorse molecules of life are proteins. Extending the analogy of DNA as a blueprint, proteins can be thought of as the engineers who convert the blueprint into reality, carrying out the instructions coded for by the DNA. Unlike the blueprints used by human engineers the DNA blueprint also has to include instructions that inform the cell how to manufacture the engineers — proteins — themselves; this is where the words written in the four-letter genetic code come in.

Proteins, like DNA, are long molecules made of smaller units. The units are called amino acids, but whereas there are four subunit types in the DNA alphabet there are 20 types of amino acid in proteins. With an alphabet of four letters only, to give a name to each amino acid requires a language of three-letter words; there are 64 different three-letter words that can be written with a four-letter alphabet, but only 16 different two-letter words. All of the three-letter words — usually called codons — have now been identified and their meaning unravelled. Sixty of the 64 refer to specific amino acids, and since there are only 20 amino acids this means that some codons refer to the same amino acid — multiple redundancy, in computer jargon. The other four are punctuation marks, a single codon that means 'start of protein message' and three different ways to say 'end of protein message'. The factories of the cell, in effect, respond to either 'halt', 'stop', or 'end', but they require a unique 'start' message before they begin work. This may be simply a fluke of redundancy, but it certainly conserves the valuable resources

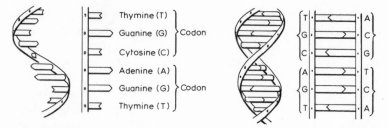

The genetic code is strung along the length of the DNA strands, which join together to make a double helix. Because bases form complementary pairs, each strand can reproduce its partner.

of the living cell by reducing the chance of it starting to manufacture a protein by mistake and having alternative commands to shut down production.

All of the information for all of the activities of life are coded in the DNA blueprints of every cell in the body, but only a few of those activities are going on at any one time. Somehow, the cell is able to read the relevant stretch of DNA needed to carry out a particular job, then de-activate it until it is required again. The stretch of DNA between the 'start' and 'stop' punctuation marks is a single gene, and codes for a single specific protein, giving rise to the fundamental epigram 'one gene, one protein'. And it is changes to the DNA message that alter the way the body works and is constructed, producing the variability within species that Darwin recognised as the key to evolution. But before we get into the intricacies of genes and evolution, it is as well to take note of the other remarkable property of DNA. What matters is not just that the DNA carries the message of the genetic code strung out as three-letter words like beads along a wire. That is useful, indeed, but what makes DNA vital is its ability to reproduce itself, copying the entire genetic code in the process. Life reproduces itself; DNA reproduces itself; DNA is the molecule of life. How does it do this trick?

Although the genetic information is stored along a single strand of DNA, the molecule is most usually found not as a single string but as a pair of twinned strands that coil about one another to form the renowned double helix. This double-stranded structure holds the answer to DNA's unique replicability, but to see how it works we must imagine the spiral staircase unwound to form a ladder. The sugar-phosphate chains now form the side-rails of the ladder, and the bases form rungs connecting the side-rails. Each rung is made of two parts, one base from each strand of the double DNA molecule, held together by so-called hydrogen bonds. The very special feature of the nucleotides is that they fit together in unique pairs. The hydrogen bonds will only form properly if thymine on one strand is paired with adenine on the other, and similarly cytosine on one strand must be opposite guanine on the other. This is because adenine and thymine each have two places where they will join while cytosine and guanine each have three; trying to pair adenine with cytosine would be like trying to put a two-pin electrical plug into a three-hole socket – it can be done, but it isn't easy, and it very seldom happens. The linked base pairs, TA and CG, each have almost identical shapes, so that the rungs on the DNA ladder are identical, and the exact way that the bases fit together and attach to the sugar-phosphate backbone explains why the strands coil into the regular helical shape made famous by James Watson and Francis Crick.

Because the bases will only pair with their complementary opposite number, A with T and C with G, it is easy to see how DNA reproduces. The two strands of the molecule separate to give two halves of a ladder that are, in a sense, mirror images. Where there is a T on one, there must be an A on the other, and where there is a C on one there will be a G on the other. Each strand thus carries all the information needed to build its partner; every unattached T will pair up with an A from the rich chemical soup inside the cell, and every unattached C will catch a G. Then, along comes an enzyme that zips the newly caught nucleotides together with a sugar-phosphate chain and that is it. Two double helices where before there was only one.

To recapitulate then, DNA codes for proteins by means of three-letter words, or codons, strung along its length; each of the codons corresponds to one of the amino acids on the protein chain. The whole DNA message contains the blueprint for the construction, care and maintenance of the individual – plant or animal – and this complete message is present in every cell. Only parts of the message are activated in any one cell at any particular time, and these active parts of the code are used to manufacture proteins which do the actual engineering work. The information on the DNA, however, not only codes for the proteins, it also codes for its own replication, because the two strands of the double helix are in fact complementary copies of each other. Because DNA can be copied, the genetic code can be passed on from parent to offspring. Those are the fundamental elements of the biochemical basis of heredity.[2]

If DNA were so well organised that whenever it reproduced it did so completely faithfully and never made any mistakes, there would be no variety at all in living things, because all would still be exactly like the first strand of DNA that managed to reproduce. But mistakes do occur, particularly during replication, and many of these mistakes affect only a single base on the DNA, a single letter of the genetic code. These are called point mutations, because they are limited to one point along the strand, and they take three forms: deletions remove one base from the strand, additions insert one base into the strand, and substitutions change one base for another. Additions and deletions are generally disastrous. A meaningful message like 'the cat and the rat sat' becomes quite garbled, reading instead 'the caa ndt her ats at '. Substitutions are less likely to be a disaster, partly because of the redundancy of the genetic code, which means that a change may have no effect at all if a codon that specifies a particular amino acid is changed to another codon for the same amino acid, and partly because even if there is a change in one amino acid it might not make a great difference to the final message, as in 'the cat and the rat sit'. By and large, though, mutations are not a good thing.

Haemoglobin, the pigment that gives blood its red colour and carries oxygen around the body, provides a good example of the disastrous results of a single nucleotide substitution. Haemoglobin is a complex molecule made up of four subunits, two so-called alpha chains and two beta chains. The alpha chains are each 141 amino acids long, and the beta chains are 146 amino acids long. Sickle-cell anaemia, a genetic disease that is invariably fatal, is the result of just one amino acid change in just one of the haemoglobin chains; at the sixth position on the beta chain there is a valine instead of a glutamine. Don't worry about the details — what matters is that this change is the result of a single point mutation in the beta haemo-globin gene, a T where there should be an A. That single, tiny difference is the difference between life and death for the person carrying it. Just one false nucleotide among thousands of millions.

Sickle-cell anaemia, as it happens, is not totally harmful because it does, in special circumstances, protect the carrier from malaria; that is why it is still relatively common in people whose ancestors evolved in malarial regions, such as the negroes of West Africa. But where the balancing pressure of malaria is removed, as it is in the United States, the advantages of the sickle cell trait disappear, and the mutant sickling gene is indeed less common among the black descendants of West African slaves than it is among Africans who never left their home continent.

Most mutations are completely detrimental. Only a very few actually improve the protein that they code for. Again, haemoglobin provides an example.[3] Whales and other diving animals are able to stay underwater a long time because they have an oxygen–carrying pigment that, compared to mammals on land, will take up more oxygen in the lungs and give up more oxygen to the tissues. Biochemical investigations have pinpointed the changes in amino acid sequence in the protein chains that make the haemoglobin more efficient and related those changes to changes in the code of haemoglobin genes. Biochemists find the same trend within the diving mammals: each has the haemoglobin that suits it, so that the blood of a deep–diving oceanic porpoise can carry three times more oxygen than the blood of a coastal one. The important point is that these changes have become part of the machinery of whales through the process of natural selection. The whale was not designed in advance for deep diving. The variations arose through chance mutations, and although most of them were as deadly as sickle cell, if not more so, a very few turned out to code for an improved haemoglobin. Ancestral whales that had more efficient haemoglobin would dive more efficiently, perhaps finding more food than those with less efficient haemoglobin or perhaps escaping from danger more effectively. Individuals with better haemoglobin were certainly at an advantage. They survived longer, had more descendants, and now the improved haemoglobin genes that at least partially gave them their success are fixed in those descendants for us to see. If the circumstances of the whales change again, so that these altered haemoglobins are no longer an advantage, then there will no longer be any selective pressure to maintain them.

Mutations, then, are generally harmful and occasionally beneficial, but sometimes have no effect, good or bad, on the reproductive success of the individual carrying it. Point mutations, however, are too subtle to have anything to do with the origin of species. That depends on altogether different mechanisms that we will discuss in more detail later. But they do have a lot to do with the details of the different proteins that different species carry, and that is why they are so useful to evolutionary biologists. Changes to the DNA will generally be resisted by natural selection, for it is very hard to improve on the past eons of evolution. But some changes are introduced every time DNA is copied, and the ones that do not have harmful effects accumulate in the DNA of successive generations. Because the DNA messages are so long, and because there are only a few ways in which point mutations can occur, errors accumulate in a statistically reliable way, like the chances of getting heads or tails when tossing a coin. Copy one codon, and there is very little point in using statistics to predict the prospect of a mistake; toss a coin three times — equivalent to the three letters of a codon — and it might well come up three heads, three tails, or some other combination. But if a perfect coin is tossed a million times, the outcome will be very close indeed to half a million heads and half a million tails. Copy a million codons of DNA, and the number of point mutation mistakes is equally reliably determined by the statistics of the copying process. Copy another million codons, and the *same number* of new mistakes are introduced. So differences in DNA accumulate as it is copied and recopied, and they accumulate at a very steady rate. When two species split off from a common stock, the ancestral forms start out with almost identical DNA blueprints. Differences then accumulate steadily in the DNA of each of the related species as it follows its own evolutionary path, and the number of differences that have accumulated in the DNA of two species is a good measure of the time that has elapsed since the ancestors of those species split from one another. All the molecular evolutionist today has to do is measure the differences — directly by reading the DNA or indirectly by examining the proteins — and relate those changes to a known point in time. Then he can not only reconstruct the evolutionary route taken by a species but also make a very good estimate of the times at which it passed each branch point.

Some well-dated fossils are necessary to establish the rate at which molecular changes occur in a particular type of protein. An unambiguous fossil date may tell us, say, that the spider monkey and the vervet monkey shared a common ancestor 35 million years ago, and that provides a calibration for the molecular clock. But once the clock has been set for a particular protein in a particular family, the molecular evolutionists are dealing with living species and can make direct measurements in the laboratory. If they need more information, they can make more measurements at any time.

How much more difficult is the palaeontologist's task. He must try to reconstruct the route from a handful of clues provided by the scattered remnants of bodies that died along the way. He cannot even be sure that descendants of those bodies ever made it to the end of the journey, and he cannot summon up new fossil evidence to plug the gaps in his knowledge. The surprise, given all the difficulties, is that they not only attempt it but so often get the story right, as far as we can tell.

Where man is concerned, the experts do not even agree on just how good the fossil evidence is, let alone the interpretation. Some palaeontologists think that the fossil evidence for human evolution is every bit as good as the evidence for any major species, like the horse, or lion. Robert Martin, who is Reader in Anthropology at University College London, has said that the human fossils are 'passably good'.[4] Others, like UCLA palaeontologist Everett Olson, feel that it is a great deal sparser. 'I take a dim view of the fossil record as a source of data,' Olson said at a recent conference on evolution.[5] Steven Stanley, another prominent evolutionary biologist, has stigmatised the hominid fossil record as being 'particularly depauperate'.[6] So what exactly do the palaeo-anthropologists have to go on? The fossil evidence of man's evolution consists of a few fragments of skull and other bones, and lots and lots of teeth. Without sugar to rot them, early man's teeth were far less likely to fall out, they were adapted to withstand the incessant wear and tear of chewing, and they contained nothing of interest — no marrow — that would attract the attention of scavengers after their owner had died. Long after most of a corpse's bones had been crunched to bits, the teeth would remain. Of course some other bones survived to become fossils, and of those we have uncovered a tiny proportion. The publicity that accompanies every new find of fossil human remains might give the impression that by now palaeontologists have a great wealth of bones to study. Not a bit of it. The known fossils of man's ancestors would all fit on a good-sized dining table, and a couple of shoeboxes would accommodate all the teeth. On this slender evidence the palaeo-anthropologists have built their theories.

This rather unpalatable truth — that there really aren't that many witnesses to man's past — is often skated over in popular accounts of how one reconstructs the history of man. Specialists who have worked with those few bones all their professional lives become so familiar with them that it is perhaps hard for them to step back from the trees and see just how small the whole wood is. To an outsider, and many insiders too, the most surprising thing is that anything at all about human evolution can be learned from the fossils. Dr Tom Kemp, curator of the University Museum at Oxford, has spent the best part of 20 years working on the evolution of the mammals. 'Anthropologists do a whole thesis on a couple of teeth that I would throw away,' he says.[7] But the truth is that they do provide a great deal of information, and have been used to construct stories that span 30 million years. Of course the stories do not all agree, and of course each one has loopholes and flaws, but given the raw material the wonder is that one can make anything of it.

The recent argument between Don Johanson, of the Cleveland Museum, and Richard Leakey, of the Kenya National Museum, illustrates the process at work. Johanson, back in 1974, discovered the most remarkable fossil, the skull and better part of the skeleton of a young female, promptly christened Lucy. Accurate datings showed that Lucy had lived some 3.8 million years ago, and her skeleton showed that, although she had a small brain, she walked fully upright. Johanson named her *Australopithecus afarensis,* after the region in Ethiopia where he found her, and redrew man's family tree to accommodate this bipedal, small-headed proto-woman. In so doing he challenged Leakey's own theories of human origins, but while Johanson sees Lucy as an absolutely central figure in the story of human evolution, and has written an entire book, *Lucy,*[8] about his find and the image she conjures up of man's origins, Leakey sees her as only a supporting actress, and devotes just five pages to her in his most recent book, *The Making of Mankind.*[9] They can't both be correct, but in the absence of better fossils it is very difficult to decide between them. That is the wonder of palaeontology — not that it is done well, but that it is done at all. (We ought also to mention, in fairness to Leakey, that Yves Coppens, of the Musée de l'Homme in Paris, disagrees with Johanson's conclusions; Coppens' opinion in all this matters not so much because he was co-leader with Johanson of the expeditions on which the contentious fossils were found, but much more because he is one of the few palaeontologists, apart from Johanson, to have studied the actual fossils rather than casts.)

The theories, whichever ones you believe, have this in common. They all try to slot new fossil finds into an existing framework until something so extraordinary turns up that it won't fit in anywhere.

Only then, with much head scratching, are the theorists willing to dismantle their favoured framework and rebuild it in such a way that the new find can be accommodated, with a place for each fossil and each fossil in its place. The frameworks are intended to show how species are related, which of them share ancestors, and which of them have left descendants. They come in two basic varieties.

Ladders, which used to be all the rage, put the hominid fossils on a single line, with each fossil's position determined by its age. The experts then looked for the changes that led from one type — one rung on the ladder — to the next, and for the 'missing links' to slot into place between the known fossils. Each fossil is supposed to give rise to a descendant who was more human, so that one climbed up the ladder from a four-legged, small-brained brute ('ape'), over a variety of stooped, shambling half-wits ('ape-men', and then 'man-apes') until one arrived at the top, where resided a fully upright being with a large brain ('man', the crown of creation). The snag with this approach — apart from the erroneous impression it creates that man marks an 'end point', an ultimate achievement that evolution has been striving towards all these eons — is that it leaves very little room for dead ends. And that makes the alternative 'bush' theories of evolution a lot more attractive than 'ladders'.

Bushes allow some fossils to lie off the direct path to man, marking the tips of twigs that stopped growing. Rather than one ladder of evolution, with its single sequence of development, we have a growing bush in which many branch tips may be reaching forward as new forms evolve and adapt, but many more branches have stopped growing and are marked by fossil dead ends back in the heart of the bush. By this reckoning, one type of early man might give rise to two types of descendant, both of which leave fossilised remains (at least a tooth or two) for us to discover. One of the lines then dies out, while the other splits again and leads, eventually, to modern man. And the whole process may have repeated many times, perhaps every 2 or 3 million years, during the 30 million years since the last days of the Oligocene. Bush theories are more accommodating than ladders because you can always invent a little branch or twig within the bush structure to provide a home for some newly discovered variety of fossil, which provides more scope to explain away odd discoveries and can, unfortunately, make it easier to cling to an outmoded framework as new discoveries are made. Even so, it is clear that bush theories are a lot more useful, and a lot more realistic, than ladders.

There were so many different hominids around in the late Pliocene, say 3 million years ago, that even the limited fossil evidence we have leaves no doubt that two or more species were alive at the same time, and often in the same place, which is just impossible on a simple ladder theory. Indeed, there were so many hominids around then that palaeontologists sometimes refer to the period, not altogether seriously it must be admitted, as the Age of Hominids, in just the same way as the Devonian era, around 400 million years ago, is the Age of Fishes, and the Cretaceous, from 135 to 65 million years ago, is the Age of Dinosaurs. Fossil evidence makes it abundantly clear that neither fishes nor dinosaurs evolved along a ladder; why should man be an exception?

Modern man is not resting on the topmost rung of a ladder, where he will enjoy a permanent place in the Sun as a reward for having climbed so far. He is, rather, the tip of one twig of a branch on a growing bush that contains other growing branches and not a few dead ends. The living branches all continue to grow outwards — evolve — in competition with one another for the light, and man's branch, like all the others, may continue to grow, or die, or fork. It certainly will not stay the same. And all of this activity can be seen as part of one great bush of life; there is no need to invent different ladders for snails, birds, insects, or wheat plants.

Bushes are now widely agreed to be a far better representation of the history of life on Earth, but we would not want to claim that the image is by any means complete or perfectly understood. Great gaps in our knowledge remain, particularly about what we might call the shape of the bush. How do branches grow and divide? What is the specific pattern of man's own bushy branch of the greater tree? How do we get from a table littered with bone scraps and a box full of teeth to a coherent theory of human origins? Full answers to these questions are beyond the scope of this book, though the last problem has been spectacularly illustrated of late by Johanson and Leakey, two of the world's leading palaeo-anthropologists. They may disagree on important points, but both give a remarkable taste of the true flavour of research in this field. As to the first two questions, we can venture some suggestions.

We have seen how DNA carries information down through the generations and how sometimes small changes to the DNA, point mutations, will give rise to an improved product that confers some sort of benefit on its owner, but we also said that mutations of this sort are not, as a rule, responsible for the origin of new species. Point mutations, plus natural selection, help to adapt a species to a particular way of life, as in the example of the diving whales and their haemoglobin. But this gradual process of evolution — often portrayed as classic Darwinism — explains better why a species should stay nicely in tune with its environment, rather than why new species should arise from old stock. The discovery that evolution of new species actually occurs through much more dramatic and sudden changes in the genetic code, almost certainly connected with dramatic and sudden changes in the environment, is sometimes presented as anti-Darwinian, but it is nothing of the sort. The new picture still includes the same three key aspects, variability, inheritance, and the struggle to survive, just as Darwin said. Darwin did not have the knowledge we have of the molecular basis of heredity and evolution, but he had collected a wealth of information showing that every animal and plant was in some sense an individual and that these individual characteristics were passed on in the course of reproduction. The origin of species, he said, took place when the small individual differences had mounted up to the extent that individuals were no longer able to interbreed freely. Most often this would occur because some geographical feature, a river or a mountain range, for example, had separated two groups of a single species. In their separate habitats there would be slight peculiarities of the environment that would favour slightly different variations from among the population. Each of the groups would become adapted to its own circumstances so that when the geographical barrier vanished, allowing the two populations to mix freely once again, they would have accumulated so many small genetic changes that they would not in fact be able to breed successfully. This theory was given a solid backing in the 1940s, when developments in our understanding of the nature of heredity and of the way populations alter under the pressures of selection led to a new formulation of Darwinism that came to be called the Modern Synthesis.

Under the new synthesis, several aspects of the origin of species were given a firm mathematical foundation, one of the predictions of which was that, as Darwin had indicated, the process of change between one species and another was extremely slow and gradual. As a result, it should be possible not only to chart the slow change of species as recorded in the fossils, but also to spot fossils that are halfway between one species and its descendants. Palaeontologists have searched diligently for these missing links in order to illustrate the gradual emergence of one species from another. For example, the giraffe is quite closely related to the deer, so there should be, somewhere in the fossil record, something that is halfway between a deer and a giraffe, with a neck and legs longer than a deer's but not as long as a modern giraffe's. There isn't.

Our portrayal of the Modern Synthesis may be a bit of a caricature, but it does capture roughly the way the old argument about speciation ran. For years the notable absence of transitional forms was an acute embarrassment to the fossil-hunters. The standard excuse, one that is still quite prevalent today, is that the missing links do exist, and are buried somewhere as fossils, but they haven't yet been unearthed. No less embarrassing, though far less widely appreciated, was the lack of any evidence for gradual change within a single species through evolutionary time. This lack, though it too tended to argue against the gradual shift of characteristics, was not a bad thing though, for it allowed the palaeontologists to use particular species as markers: if one found some species in a rock one could be more or less sure that the rock was of the same sort of age as other rocks in which one found the same assemblage of species, and this was very useful for dating certain types of rock that couldn't be dated any other way. So, while the palaeontologists were content, and able, to make use of the fact that species did not apparently change during their history, they couldn't really admit it. And they also were put in something of a bind because they had to explain publicly the distinct lack of in-between specimens by blaming it on the vagaries of the fossil record. The palaeontologists were, without knowing it, echoing the sentiments expressed by Cambridge astronomer Martin Rees, uttered in another context. At a gathering during a high-powered international seminar to discuss the 'problem' of the Universe's missing mass — we can only detect some 10 per cent of the mass needed to keep the Universe from expanding for ever — some people were eager to say that the missing mass does not exist. Rees, to a chorus of quiet academic chuckles, reminded them that 'absence of evidence is not evidence of absence'.[10]

Perhaps, though, Rees and the palaeontologists are wrong, and the fossil record is accurate. Darwinism may be correct, but the Modern Synthesis may be wrong. That essentially was the thought that led Stephen Jay Gould and Niles Eldredge to develop their theory of punctuated equilibrium.[11] As Gould says, 'certainly the record is poor, but the jerkiness you see is not the result of gaps, it is the consequence of the jerky mode of evolutionary change'.[12] Evolution jerky? What about Darwin's smooth accumulations of minute changes?

The newest ideas on speciation almost make a virtue of the missing missing links, and use their absence as positive evidence in support of the idea Gould called punctuated equilibrium. This says simply that there is no gradual change within a species, except possibly in size, and that species are remarkably constant over long periods of time. What keeps them constant is, of course, natural selection, weeding out deviant forms. Gould describes the evidence thus: 'For millions of years species remain unchanged in the fossil record and they then abruptly disappear, to be replaced by something that is substantially different but clearly related.'[13] Long periods of equilibrium are punctuated by bursts of very rapid change, and the chances of one of the transitional forms being fossilised is very slim indeed. Where subscribers to the Modern Synthesis, often called gradualists, picture one species giving rise to another over hundreds of thousands of years, punctuationists believe that, depending on the mutation involved, the process could occur within a single generation, but that it might take 10,000 years or even longer. Long though that might seem, it is all but invisible measured against the huge stretches of geological time (65 million years since the death of the dinosaurs alone), and the chances of picking up a fossil from the short period of change, as compared with the much longer period of stability, are remote indeed.

The chance of finding a missing link if the change occurred literally between one generation and the next is, of course, zero, and some punctuationists argue that some evolutionary jumps (although not from deer to giraffe) really happen that quickly. But the theory of punctuated equilibrium does give a very good idea of how to go about searching for a missing link, even though it might have existed for only a short time. You need to look at an animal that existed in vast populations for a relatively long time and which was very likely to become fossilised. Molluscs inside their shells provide an extremely good candidate, and Peter Williamson, a colleague of Gould's, spent a lot of time searching the fossil beds around Lake Turkana, where Richard Leakey's team looks for ancient man, examining the fossilised snails there. He found one species that had the usual rounded edge to its shell, and another that was very similar except that the edge of its shell was more angular, like a helter-skelter. The rounded shell occupied the fossil beds in huge numbers and persisted, unchanged, for millions of years. Then, quite suddenly, it gave way to the shell with the triangular edge, in the same vast quantities. Williamson examined the geological boundary between the two species, and there, among the normal shells, he found a population of snails that were clearly in a state of transition, part rounded and part triangular. Genuine missing links, whose existence and discovery was predicted by the very theory that seeks to account for the absence of missing links.[14] Absence of evidence is not evidence of absence.

We will return to the fascinating details of punctuated equilibria later, but having talked gaily of species and speciation, and of one species evolving into another, we should perhaps make it clear that the species can be a very slippery concept. In general, the specialist uses the word to cover a type of plant or animal that people recognise as different from others and grace with a name. The blackbird is a species, and a different species from the songthrush or the mistle thrush; the daffodil is a different species from the narcissus. Interestingly, this correspondence of a biological species with a common name applies equally well whether the species concerned are named by an Englishman in his garden or a New Guinean in the deep forest; wherever they are, people tend to give different species different names. Clearly, however, there are many more species than have been christened by people, and the question of definition is more complicated than simply naming things that look different.

The taxonomist really has two possibilities. If he has a large number of specimens he can use a sort of mathematical method, defining species by saying that the variation between members of the same species is smaller than the variation between members of different species. This can be a very fruitful method, but it is also subject to personal and often arbitrary decisions. A great dane and a chihuahua, for example, are members of the same species, *Canis familiaris*, but they are grossly different, much more so than members of a species are naturally. Alternatively, the taxonomist can go for a more functional approach, using what he knows of the specimen's biology. The best biological definition of a species is that two organisms are members of the same species if they are able to

THE EVOLUTIONARY CLOCKWORK

breed and have fertile offspring. A horse and a donkey can breed, and the offspring of their union is called a mule, but the mule is sterile and cannot breed, so the horse and donkey are regarded as separate, but closely related, species. An arab stallion and a carthorse — superficially every bit as different as the horse and the donkey — produce a foal that is fertile and can breed itself: they are members of the same species. (Obviously a little common sense is required in using this definition. We wouldn't want to say that two men, or for that matter two women, are members of different species just because they cannot breed.)

This definition of species is a rough and ready one, useful in most circumstances but occasionally confused by exceptions that we will not go into here, although it is worth acknowledging that if some new species are evolving all the time, as they surely are, then we would expect to find occasional confusing cases in which speciation is in progress and our working definition is, in fact, unworkable. There is a far more serious problem with the definition of species in terms of breeding that faces the palaeontologist. The living biologist, for want of a better phrase (some call themselves neontologists, to distinguish themselves from palaeontologists), can always look closely at his animals and even do experiments to find out whether they are one species or several, but the fossil biologist — palaeontologist — has no such luck. Fossils don't breed, so how can he ever decide whether two similar-looking fossils are members of the same or different species? Even the abrupt changes predicted by punctuated equilibrium may be physically rather small and hard to spot, and without a large number of specimens it becomes very hard to use mathematical methods. So how is one to decide when one species has changed into another?

The palaeontologist takes a pragmatic approach, tracking his species through time and giving it a new name if it seems to have changed noticeably. The portions between changes are generally called species, but they should properly be called chronospecies; a chronospecies is a segment of a lineage, which is a sequence of biological species related to one another by descent.

The distinctions between the chronospecies that comprise the lineage may be obvious, as for example are the differences between *Homo habilis* (whose story we tell in the next chapter) and *Homo erectus*. Equally, they may be very subtle or even non-existent, and the temptation to make the most of small differences so that one can

(51)

create a new species and achieve scientific immortality by choosing a name for one's fossils is very strong indeed. Ever since the original Neanderthal man was uncovered in 1857 there have been hordes of distinct species described for hominid fossils. Many of these have now fallen by the wayside, but even today the arguments continue. Is Don Johanson's Lucy the same species that made the evocative footprints discovered a thousand miles away at Laetoli by Mary Leakey? We certainly don't know, and we doubt that they know either; with so few examples from the human family tree it is a very wise man − or a fool − who claims to be able to distinguish the 'normal' variation that characterises members of one species from the 'abnormal' variation that sets species apart.

However one thinks of a species, there are two things that can happen to it in the course of evolution. First, it can become extinct, leaving no descendants at all. Second, it can split to give rise to a new species. Those are the two basic alternatives, but as we've already seen palaeontologists may find that a single lineage recovered from the fossil record shows sufficient change to merit being divided into a number of chronospecies. Often there is a strong assumption that the species do indeed come from a lineage rather than a series of related branches intercepted at different times, but even where this is justified the fact is that a chronospecies does not really go extinct. It is replaced by the next chronospecies in the lineage, and extinction should refer only to the end of a lineage, just as speciation should refer only to the splitting of a lineage.

Even if species do evolve suddenly, as it seems they do, we are still faced with the same problem that faced Darwin and all of his successors. What are the mechanisms behind this abrupt shift? Slow accumulations of point mutations would seem to have been ruled out, but what does that leave us with? 'Not a lot' must be the honest answer, but there are some suggestions. One is a process known as chromosomal evolution, and this is particularly relevant to human evolution.

DNA does not float about freely in the cell (except in the cells of bacteria). It is packaged with certain types of proteins into rod-shaped bodies called chromosomes, and the structure of the chromosome is at least partially responsible for controlling the everyday working of the genes. Every body cell of a particular species has a characteristic number of chromosomes, usually called the chromosome complement. Fruitflies each have 8, while man has 46 and chimps have 48. But don't feel superior; some ferns have more than 600 chromosomes in every cell. Chromosomes reproduce in essentially the same way as DNA, with double strands separating and making copies of themselves, but there is a class of mutations that affects not points along the DNA but whole stretches of it along the chromosome. These are called chromosomal mutations, but because it is the DNA that is affected we will describe them in terms of DNA. Sometimes a length of DNA will become detached from one chromosome and attached to another. This is called a translocation, and mutations of this sort have been identified and associated with a plethora of genetic diseases in man. Alternatively a bit of the DNA might be snipped out of the whole strand, turned upside down, and reinserted into the chromosome. This type of mutation is called an inversion, and it bedevils certain typesetters, who are forever mutating 'CAUSAL' into 'CASUAL', generally with lamentable results.

As with point mutations, most chromosomal mutations will be disastrous, but occasionally one will have beneficial consequences, and the likelihood is that those consequences will be much more wide-ranging than a simple point mutation's. This is because of the way genes are controlled, and the importance of that control in development. Now the truth is that we know very little about how genes are switched on and off in particular cells, and even less about how the information on the DNA is translated into the instructions that shape a formless mass of cells into an embryo that develops into a properly formed adult. But what we do know suggests that there are two classes of genes, structural genes and regulators. Most genes are structural; they code for proteins that are needed and used by the cell, either as engineers or as raw materials. A few are regulators, so-called because they code for proteins whose function is to regulate the switching of the structural genes. This distinction is a very important one, to which we will return, but the primary aspect of it as far as chromosomal mutations are concerned is that these shufflings of the DNA can have the effect of bringing a particular structural gene under the influence of a new and different regulator. This may switch it on at different times, or for longer periods, or whatever, and in doing so may produce an animal that is grossly unlike the ancestor without the mutation. A change in one small stretch of DNA can thereby produce a sudden and dramatic change in the

(53)

body built from the whole blueprint.

One of the regulators that is known, or rather inferred, controls the development of the bones in a chicken's leg. The limbs of all vertebrates above the fishes are built on the same sort of plan, the pentadactyl limb, but with modifications. Horses, for example, have only one, enormously elongated, toe, instead of the usual five. And chickens, like other birds, have a very reduced fibula, the smaller bone alongside the shin bone, as you will know if you have ever paid attention to eating a drumstick. Along with the reduced fibula the chicken also has many of its ankle bones missing, though this wouldn't be so obvious to the casual chicken-leg eater. It turns out that, by a variety of surgical manipulations on the chicken fetus in the egg, one can persuade a developing chick to grow a much sturdier fibula and also develop some of the normally missing ankle bones. Chicken DNA contains the information to build the 'normal' limb; so why doesn't it? The smart money is betting that it will turn out to be due to a very slight change in a regulator gene. The mutants never grow an ankle without a fibula, but once the fibula contacts the ankle region the bones there develop. The interpretation of this is that the presence of the fibula somehow induces the ankle bones to form, and that at some stage in its past the structural genes for the chicken's leg came under the influence of a new regulator that controlled them in a different way. The effect was to reduce the growth of the fibula, which in turn modified the ankle. The structural genes themselves are still there, as the experimental birds show, and what happened in evolution was that they got switched off.[15]

Chromosomal mutations, then, can have a powerful effect, and what is more they crop up in the space of one generation. The major problem that they pose for the animal carrying them is that it will surely have a very hard time finding a mate. The cellular mechanics of sexual reproduction mean that matching chromosomes from the male and female have to pair up before development can proceed. If one of the chromosomes contains a translocation or an inversion it is unlikely to be able to pair up with its usual partner. But if the social system of an animal involves a single dominant male copulating with a number of females in a harem, then a chromosomal mutation that arises in the male could easily be distributed among a large number of offspring, some of whom might eventually find one another and mate. In this way the chromosomal mutant could quickly become fixed in a small population and its presence would be an effective barrier to mating with the parent species. Even if it did not make a great difference to its carriers it would have driven a small wedge between them and the rest of the species so that the new species would now be forced to evolve separately.

Guy Bush, a geneticist at Texas University who is working on models of speciation that rely on chromosomal rearrangements of this type, is confident that the system is not as unlikely as it sounds. He also points out that 'favorable social organizations are relatively common, in horses, many primates, and rodents, for instance'.[16] Is it just a coincidence that he mentions primates? Aside from the notable similarity of their DNA at the level of bases, man, chimp and gorilla also have extremely similar chromosomes. Just six inversions distinguish the chromosomes of man and chimpanzee, six mutations that maketh man.[17]

Other mechanisms that would promote the rapid change of lineages required by punctuated equilibrium theory have been proposed, but it is early days yet and most biologists are still having a hard time coming to terms with the very idea of punctuated equilibrium. The problem is one of scale. The conventional version of Darwin's theory, as promulgated in the Modern Synthesis, emphasised the uniformity of life processes. What went on at higher levels of organisation, such as the species, was supposed to be fundamentally the same as what went on at the level of the individual, and just as members of a species could adapt to their environment, so the species could change and adapt too. Part of the opposition to punctuated equilibrium has been based on a reluctance to accept different mechanisms at different levels of organisation, but a good deal of it also stems from inertia. Having accepted Darwin, most people are content to have his ideas remain for all time. That is why suggestions are repeatedly made that the idea of punctuated evolution indicates that Darwin was in some way wrong. This supposition is incorrect, for two reasons.

First, punctuated equilibrium does not deny the relevance of natural selection to either the individual or the species. Natural selection still operates to fit a species to its environment, with adaptations that track the environmental fluctuations but average out over a longer timescale to provide a picture of unchanging stasis. It also operates to decide which species live long and give rise to many branches and which become extinct. Speciation is still followed by selection, regardless of the time span over which the speciation takes place. Whatever the reason for the shape of the modern chicken leg, modern chickens are around today because that leg structure has been proved by natural selection to be an advantage to the chicken's way of life.

Secondly, Darwin himself was not the out and out gradualist that he is remembered as. For example, in all editions of *The Origin of Species* after the third, he wrote 'the periods during which species have been undergoing modification, though very long as measured by years, have probably been short in comparison with the periods during which these same species remained without undergoing any change'.[18] One could hardly wish for a better statement of the punctuationist position.

Naturally the debate over the ways in which species arise is still continuing, but there does seem to be a groundswell in favour of punctuated equilibria. There is a great deal of evidence that particular lineages in the fossil record do not show any gradual change at all, and this evidence is now so good that Francisco Ayala, a population geneticist and one of the major proponents of the gradualistic Modern Synthesis, has admitted that he is 'now convinced from what the paleontologists say that small changes do not accumulate'.[19]

Perhaps we should not be surprised that one lineage over which the debate is particularly protracted is our own. Man has always been held out as one of the better examples of gradualism, slowly developing his large brain and upright stance over a long period of time. Moreover, the gradualist position maintained that no more than one hominid species ever existed at one time, so that the development of the lineage was one of successive chronospecies. This has now been shown to be false by the very good evidence of different species of hominid living not only at the same time but also at the same places. It now seems that there may have been four species of hominid all alive at the same time, about 2 million years ago, but even this is not the strongest evidence against gradualism in our ancestry. Steven Stanley, who works at Johns Hopkins University, has looked at how long each of the various hominid species lasted, and compared them with other African mammals of the same time. He finds that on average the hominid species have lasted just as long, or even slightly longer, than the species of other mammals. To put it another way, the rate of speciation in our own line is a little below the rate in, say, the giraffe line. 'This,' as Stanley says, 'is a rather startling revelation, considering the extraordinarily high net rate of evolution that is the hallmark of human [evolution]'.[20] What he means is that because the species themselves are constant, gradualists can only account for the massive physical changes if they take refuge in completely undiscovered chronospecies — whole lineages of missing links — in which man's ancestors evolved far more rapidly than in the chronospecies we presently know about. 'This could hardly represent a comfortable refuge,' says Stanley, and 'we seem forced to adopt the punctuational view that distinct lineages of hominids evolved rapidly by way of small populations. In short, all evidence suggests that the traditional view of hominid evolution is wrong.'[21] Stanley's conclusion is in direct opposition to the received wisdom, for he says that evolution in each of the known hominid species has been slower, not faster, than average. The leaps taken in jumping from one species to another, however, have been large, and that is what gives the impression of rapid change.

In the easy cases, where all the evidence is available, the palaeontologist's task is really no more difficult than the neontologist's, and there has been a great deal of fascinating research in the history of life on Earth, all founded firmly in the fossil record. It is in the difficult cases, where there is not enough evidence, that the living biologist is at such an advantage. If the relationships of two types of animal confuse him, he can apply a whole battery of additional tests to see whether they are an exception to the general rules or not. The palaeontologist has no such luck; faced with a confusing picture he can hope to find further fossils that will clear up his confusion or he can try to squeeze even more information out of the fossils he has got. Or, if he really wants an answer, he can look up from his fossils and search for other sources of enlightenment.

That said, we must admit that the history of palaeontology does not read as a shining example of the pursuit of truth, especially where it was the truth of man's origins that was at issue. From the very beginning, when limestone quarrymen uncovered the skull of a man in a cave high above the Düssel river in the Neander valley, to the recent past, when Don Johanson uncovered a 'family' of early people in the high hills of Ethiopia, controversy has surrounded almost every discovery and, more especially, its interpretation.

The battle between evolutionists and their opponents over the interpretation of the Neanderthal fossils raged long and hard, with the most extreme anti-evolutionist position championed by F.

Mayer, professor of anatomy at the University of Bonn. He said that the Neanderthal specimen was the remains of an ailing Cossack soldier, a member of General Tchnernitcheff's army which had camped near the cave before it advanced across the Rhine on 14 January 1814. Mayer believed that the Cossack, probably afflicted with rickets and very ill, deserted the army, hid in the cave, and died there. Thomas Henry Huxley 'noted that Professor Mayer had failed to explain how the dying man had managed to climb a precipice twenty metres high and bury himself after his death; and wondered why the man would have removed all his clothes before performing these wonderful feats'.[22] A second skull, found during excavations in Gibraltar some time before 1848, came to light in the fuss over Neanderthal man. Practically identical to the Neanderthal specimen, the Gibraltar find added greatly to the weight of the evolutionists' arguments; as one of them, George Busk, put it, 'even Professor Mayer will hardly suppose that a ricketty Cossack engaged in the campaign of 1814 had crept into a sealed fissure in the Rock of Gibraltar'.[23]

The same general story is repeated throughout the history of palaeo-anthropology, like a bad joke that refuses to vanish. The luminaries of British palaeontology fell for a rather crude hoax because it happened to fit exactly with what they expected; other palaeontologists who were not as committed to the missing link it allegedly represented were not as readily taken in by Piltdown man. Lord Zuckerman proved by calculations, later shown to be incorrect, that *Australopithecus* was an ape, not a hominid. Raymond Dart took the bones accumulated in South African caves, smashed by carnivores and by the very process of falling into the caves, and created an 'osteodontokeratic culture' (meaning bones, teeth and horns) in which violence and aggression were rife. Don Johanson finds two sorts of fossil, of very different size, and puts them into one species, arguing that the males are bigger than the females; yet the magnitude of that size difference is greater even than is found in present-day gorillas.

Stories like these are legion in palaeontology, and we can do no better than to refer you to John Reader's fascinating book *Missing I inks*, in which he describes the chequered histories of 12 important fossils. Reader concludes that 'preconceived notions have played a fundamental role in the study of fossil man', because 'fossils are often so broken, distorted or incomplete that different authorities may stress different features with equal validity, and the points distinguishing their interpretations may be so slight or unclear that each depends as much upon the proponents' preconceived notions as upon the evidence of the fossil.' As if that weren't enough, Reader also notes that 'it is remarkable how often the first interpretations of new evidence have confirmed the preconceptions of its discoverer'.[24] Bearing all this in mind then, we now present the traditional story of man's ancestry, a story based on faith, hope, and a table-full of fossil scraps.

Three
THE TRADITIONAL TIMESCALE: GRANDFATHER'S CLOCK

Mankind is a very recent addition to the inhabitants of the Earth. Our planet has been in existence for some 4½ thousand million years; life has existed in the oceans for perhaps 3½ thousand million years, and on land for more than 300 million years. These sorts of time periods are very hard to imagine. Just how long is 4½ thousand million years anyway? The customary way to make these vast stretches of time more manageable is to produce an analogy that is a little more relevant to our daily existence. One such reduces the history of the Earth to a day that begins just after midnight, in which case the amphibians crawled out on land at 8 minutes after ten the next evening, the dinosaurs went extinct at 21 minutes to midnight, man's ancestors split from the apes at 6½ minutes to midnight, and modern man appeared less than one second before midnight. But even this analogy makes it hard to comprehend just how vast are the stretches of geological time, and in any case this way of portraying the history of the Earth, or one very like it, has been worked to death. So what about this: imagine that your outstretched arms represent the history of the Earth. The Cambrian, when life really begins to burgeon, starts somewhere in your wrist and the massive Permian extinction, 225 million years ago, is roughly at the end of your palm. The entire Cenozoic, from the death of the dinosaurs to the present day, is but a fingerprint, and man's reign almost imperceptible. John McPhee, who developed this particular analogy in his book *Basin and Range,* says baldly, 'in a single stroke with a medium-grained nail file you could eradicate human history'.[1] Geological time takes quite a bit of getting used to.

From the time the first amphibians moved up on the land some 350 million years ago until the end of the Mesozoic Era about 65 million years ago, the evolution of land animals seemed to be following a definite line. The amphibians were replaced as the dominant large animals by reptiles, and 200 million years ago one group of reptiles, the dinosaurs, achieved a position of dominance that they held on to for nearly 150 million years. Small mammals, including the first primates, appeared while the dinosaurs ruled the Earth, but they could not compete directly with the well-established dinosaurs, which by then occupied all the ecological niches suitable for large animals. The dinosaur equivalents of lions, elephants and gazelles allowed no room for mammals to evolve into genuine lions, elephants and gazelles, and the way was not clear until about 65 million years ago, when the world was transformed by a huge catastrophe that killed off all the large animal species, including all the dinosaurs. Very few species survived, and those that did were generally rather on the small side. These survivors included the ancestors of the mammals, who had been living in the shadow — and probably under the feet — of the dinosaurs. The death of the dinosaurs made available a whole host of empty ecological niches which presented ripe opportunities for new large animals. New large animals did evolve to fill those gaps, and the stock they evolved from was that of the small mammals that had hitherto been kept down. From a world of dinosaurs, the Earth became, very rapidly, a world of mammals.

What happened to kill off the dinosaurs so abruptly? We don't really know for sure, but there is a very strong possibility that disaster struck from space — from beyond the Earth and possibly even beyond our Solar System. The dinosaurs died out at the end of the Cretaceous, 65 million years ago. Rock strata of the same age have been found to contain an unusually high proportion of the element iridium, which occurs in only minute traces on the surface of the Earth. The implication is that the iridium in the 65-million-year-old rocks is not from the Earth but arrived suddenly from somewhere else. Iridium is not common on Earth, but it is relatively more plentiful inside asteroids, chunks of rocks in space that never quite made the grade as planets; one possibility is that a massive meteorite, a fragment of an asteroid, slammed into the Earth with a blow powerful enough to send vast clouds of dust and steam into the high atmosphere. The dust clouds blocked off the light from the Sun, which killed off the plants on which large animals depended for food. It also reflected away the Sun's heat, setting in train a cold epoch, or perhaps even a full ice age, which finished off those few dinosaurs that had lingered on.

Alternatively, the iridium may have come not from a meteorite but from the debris of an exploding star, a supernova, which sent massive doses of radiation sleeting across space and over the Earth. The cosmic rays might have affected the larger animals directly, or they might have destroyed the Earth's protective ozone layer. This would have allowed ultra-violet radiation from the Sun down to the surface of the Earth, where again the plants would suffer, leaving nothing for the large herbivores to eat and no herbivores for the carnivores.

Disaster from space is a common enough science-fiction theme, but in the case of the Earth and the great dinosaur death there is very good evidence to support it. Large sea creatures, for example, were not affected by the dramatic changes at the end of the Cretaceous, but air-breathing aquatic reptiles were. This is just what we would expect if disaster hit from above. And it is likely that one day, when man returns to the Moon and studies its surface in more detail, the evidence of a cosmic catastrophe 65 million years ago will be un-ravelled and the puzzle resolved. If the surface of the Moon also contains additional iridium in 65-million-year-old rocks, then the disaster was an exploding supernova; if there is no iridium layer on the Moon, then the disaster was a meteorite that struck only Earth. Meanwhile, the event, whatever it was, provides a convenient marker, a boundary in geological time that signals the end of the Mesozoic and the start of the age of mammals, the Cenozoic Era. And intriguing through the puzzle is, the reasons for the death of the dinosaurs are less important to the story of human evolution than the fact that the dinosaurs did die out.

Sixty-five million years, the duration of the age of mammals so far, is a long time, long enough for the geography of our planet to have changed drastically as continents broke up and drifted apart. And yet it represents less than 1½ per cent of the history of the Earth, scarcely an eyeblink. We are very much aware of this enormous background, almost 99 per cent of the whole, to the story we have to tell, but we must focus on the story of man on Earth. The new picture, as we've said, shortens man's tenure even more, but even the traditional timescale really only takes up the story some 30 million years ago, and before we return to the new discoveries we should re-tell the story of man's ancestry as it has been constructed from the fossils. We have said that we believe this story to be wrong on several important points, but this is not entirely the fault of the scientists who made up the story. They developed an interpretation based on the best evidence they had available, fossils. The problem is that fossils are sparse and unreliable.

For bones to be preserved as fossils, and to survive so that we may find them millions of years later, their owners have to die in rather special circumstances. For one thing, a good fossil should avoid being eaten. Most of our ancestors were probably eaten by the hunters and scavengers with whom they shared the Cenozoic scene, but even so some very important fossils did not entirely escape the attentions of scavengers: the type specimen of *Homo habilis*, the one that earned Handy Man his name, has a clear groove from a predator's tooth on his skull. The best thing for a potential fossil to do is to lie down quietly to die on the banks of a lake or river where floodwaters can deposit silt and mud over the remains. Then, after the soft parts have rotted away, chemical changes begin to transform the bones into a rocky mixture, set hard within the matrix of different rocks formed by the mud and sediments. Layer upon layer of rock can be built up in this way, each one containing its fossil remains.

The rocks hold their fossils, of that we can be sure, but we can only find the fossils when the rocks are exposed to wind and weather, steadily eroding them away and exposing the fossils. That is why so many good fossil hunting grounds are in inhospitable deserts; the climate there ensures maximum erosion, with occasional heavy rains doing the work of thousands of diggers and the absence of plant life ensuring that there is no great build-up of soil to slow the process. Chance plays a large part in deciding which bones are preserved in the first place, and an equally large part in deciding which of those fossils become exposed for our observation. Nor do fossils come neatly packaged as whole skeletons. More likely there are just fragments, and a shinbone here, a jawbone there, and piece of skull from somewhere else have all to be considered together. Do they belong to the same individual? The same species? Are they even of the same age? Can they be used to provide a reliable picture of what some particular ancestor of ours looked like? Add to these problems the distortion that many tonnes of overlying rock can induce in a fossil, and you begin to appreciate that fossil-hunters need not only luck to find the fossils but a great deal of patience and an equally large amount of imagination if they are to clothe the scraps of bone in flesh. The real wonder, as we've suggested before, is that the fossil evidence is sufficient to give us any idea at all of the course of human evolution.

Such evidence as there is from the very beginning of the Cenozoic tells us that the primates inheriting some of the ecological niches vacated by the dinosaurs were small, squirrel-like creatures, which already possessed many important features that were to play a part in the emergence of man. These included eyes on the front of the face rather than at the side, paws able to grasp, and a brain relatively large for their body size, each of which is best understood as an adaptation to life in the trees. Here good eyesight, especially the stereoscopic vision provided by forward-facing eyes, good hands, and an alert brain, are essential.

Relatively soon after these early primates appeared on the scene their lineage split; one side of the split is still with us today in a fairly primitive state, and these are the lemurs, lorises and tarsiers, the so-called prosimians. The other side of the split is also still with us, and indeed is the focus of the story, for it gave rise to the monkeys which in turn gave rise to the modern apes and ourselves. Most of what we know of the earliest monkeys and apes comes from a set of fossil remains preserved at just one place on Earth, the Fayum depression on the eastern edge of the Sahara desert in Egypt. The fossils from that site provide a great deal of knowledge about the early days of the primates, but aside from the Fayum fossils we can also learn a great deal, not only about primate evolution but also about the history of the Earth, by comparing the two living families of monkeys, the New World monkeys of South and Central America and the Old World monkeys of Africa and Asia.

The history of the Earth enters into the story because of the way the great land masses shifted position during the Cenozoic. Continental drift is now a well established fact which explains how the Earth came to have its present arrangement of continents and oceans. Indeed, such is the sophistication of modern tape-measures — laser rangefinders and satellite-borne surveying equipment — that it is now possible to measure continental drift directly, in the widening of the Atlantic Ocean by a couple of centimetres each year. Continental drift is not an entirely steady process, but tends to happen in bouts. The current round of continental drift began about 200 million years ago, as the dinosaurs were coming into the ascendant and at a time when all the land mass of our planet was congregated into one giant supercontinent, Pangea. The first split was into two lesser supercontinents, Laurasia in the north and Gondwanaland in the south. These then broke apart into the continents we know today,

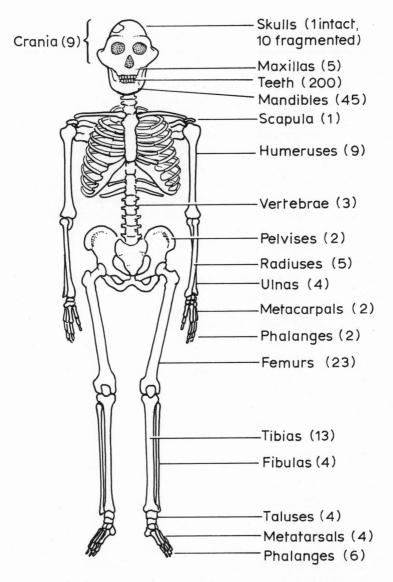

Crania (9) { ——— Skulls (1 intact, 10 fragmented)
——— Maxillas (5)
——— Teeth (200)
——— Mandibles (45)
——— Scapula (1)
——— Humeruses (9)
——— Vertebrae (3)
——— Pelvises (2)
——— Radiuses (5)
——— Ulnas (4)
——— Metacarpals (2)
——— Phalanges (2)
——— Femurs (23)
——— Tibias (13)
——— Fibulas (4)
——— Taluses (4)
——— Metatarsals (4)
——— Phalanges (6)

Reviewing the fossil finds from Lake Turkana in 1978, Richard Leakey drew up this skeleton to indicate the numbers of each type of bone. Note the large number of teeth, and scarcity of just about everything else. (Modified from *Scientific American,* August 1978, p 56.)

drifting over tens of millions of years into their present positions on the globe. Probably the last major break-up in this whole process was when the remnant of Gondwanaland split apart into South America and Africa, beginning about 40 million years ago. It has taken those 40 million years for the two continents to separate and give rise to the South Atlantic Ocean between them, and during that time the plants and animals carried on the two land masses have all followed their own separate evolutionary paths. Despite the separation of the two continents, and the separate development of their inhabitants, we can be sure that those inhabitants once shared a common ancestor. In particular the monkeys of the New and Old Worlds must be descended from the prosimians that ranged across the double continent during the 20 million years from the demise of the dinosaurs until the final break-up of the last of Gondwanaland.

The two monkey families provide a striking example of the process called parallel evolution. Starting from the same basic prosimian stock, and with the same opportunities to make a life for themselves in the trees, the evolutionary pressures of an essentially similar environment have produced essentially similar animals. The most obvious difference between the two families is in their tails — Old World monkeys use their tails just for balance, but some New World monkeys have evolved a tail that is prehensile, able to grip branches as effectively as a fifth hand. There are other differences too, in their teeth and the position of their nostrils, but perhaps the most important difference between the two monkey families is that the New World monkeys stayed monkeys. In the Old World, and only in the Old World, the tree branched again to produce another family, the apes. As Darwin guessed, the cradle of mankind was Africa, and the earliest ape remains yet discovered were found at that unique site in the Fayum depression.

Forty million years ago the now arid basin of the Fayum was covered by thick tropical forest, traversed by rivers making their way slowly north to the Mediterranean sea. The forest supported, among many other animals, monkeys and early apes, and some of these died and fell into the rivers, where mud covered their bones and began the long business of fossilisation. If the Fayum were still tropical forest it is most unlikely that we would ever find those fossils, but because it has become bare desert the fossils are exposed in the eroding sandstone. One key fossil from the region marks the earliest known specimen of our own ape line. It is an almost complete skull — extremely rare for anything so old — with an age of 28 million years. Clearly not a monkey, this skull is the first ape; in honour of the country where it was found the species has been named *Aegyptopithecus* (Egyptian ape).

Aegyptopithecus is often made to occupy the bottom rung on the ladder that leads to man; it is said to be the oldest ape ancestor. However, the fossil itself suggests that in a literal sense this is impossible, for it is almost certainly not an adult specimen. The crowns of the cheek teeth are hardly worn, and the large canines still have some way to erupt, both factors that suggest that the animal was not fully mature. As Elwyn Simons, discoverer of *Aegyptopithecus*, admits, 'this individual, probably a male, may never have mated, and if this is so he would not be anything's ancestor'.[2]

There is no doubt that the modern apes and modern man share *Aegyptopithecus* as a common ancestor; the doubt lies only in the question of when the line descending from *Aegyptopithecus* branched into the lines that produced the various apes alive today. Nor is there any real doubt about the age of these fossils. A date such as '28 million years ago' is the best date the palaeontologists can assign, and of course really means 'about 28 million years ago'. No one would be very surprised if some radical improvement in the technology used to date fossils produced a new date for the *Aegyptopithecus* skull of 30 million years ago, or 27 million years ago, say. But even that relatively small spread of doubt still qualifies it for the title of oldest known ancestor of the human and ape family. The important point about dating is not that the accuracy is some fixed amount, in the case of *Aegyptopithecus* about 3 million years, but that it is a percentage of the actual age of the fossil. So a younger fossil can be dated with a seemingly greater degree of accuracy, although of course the accuracy is a fixed proportion of the 'true' age of the fossil. A skull dated as 5 million years old, for example, might actually be anything from 4½ to 5½ million years old, while the date of a fossil from 100 thousand years ago can be ascertained with an accuracy within 10 thousand years. The error in the dating is always some small fraction of the actual age, so that the position of most remains in the hierarchy of descent from *Aegyptopithecus* can be sorted into a sensible chronology. A chronology, however is not the same thing as an evolutionary tree, and very interesting problems arise when we find more than one species alive at the same time; which is the dead-end and which the growing branch?

The techniques of dating vary from one site to the next, and the most limiting thing about all the techniques is that there is no way of dating the fossils themselves. The famous radiocarbon, or carbon-14, technique will reveal the age of any organic remains, including bone and wood, but it is no good for the fossils of man's long-distant history. This is because it works by measuring the amount of radioactive carbon in the sample, and radioactive carbon decays relatively quickly, so that if the sample is older than about 50,000 years there is too little radioactivity to measure. For fossils older than this, the palaeontologist relies on dating the rocks in which the fossil was embedded, using geological information about the rock strata. The techniques vary, but one crucially important calendar is provided by another radioactive isotope, potassium–40. The fact that potassium–40 occurs widely in volcanic ash, and volcanic ash often covers fossils, is particularly useful. The amount of potassium–40 left in a volcanic ash deposit declines slowly with time as the isotope decays into another element, the inert gas argon–40, which means that scientists can date volcanic ashes by measuring their potassium–40 content; the more of it there is, the younger the ash. The principle is exactly like the carbon-14 clock, except that the potassium–40 clock extends much further back into the past as a result of the much slower decay of potassium–40. There are other techniques for dating rocks, but for fossil remains a few million years old the potassium/argon method has the twin advantages of simplicity and reliability, and this means that to our earlier list of desirable features for fossil formation, and finding, we should add the presence of active volcanoes. From time to time these will spew out ash that will carpet the ground and become incorporated into the rocks, and as each layer of rock can be dated we have a much better chance of being able to date accurately the fossils we find if they are sandwiched on either side by layers of volcanic ash. If a fossil has a layer of ash 2 million years old above it, and another layer 2½ million years old below it, we can date the fossil with an accuracy of half a million years, and an active volcano will produce ash layers very much more often than this, which makes the whole business even more accurate.

Finally, perhaps, we should add the most important condition. As well as rivers and lakes bearing silt, and volcanoes spewing out ash, and scavengers ignoring the bones, and everything else, you need to have hominid ancestors living, and dying, in the region. It is hardly surprising that accurately dated remains of our ancestors are so scarce. The figures are instructive. Kay Behrensmeyer, who has worked extensively at Richard Leakey's camp at Koobi Fora in Northern Kenya, estimates that over a span of 1½ million years only 4 in 10,000 hominids became fossilised. And at Omo in Ethiopia, which has yielded more fossil ancestors of man than any other area, the 215 known individuals represent just 3 fossils for every million hominids.[3] Fossilisation is a very chancy business.

The best sites for hunting down our ancestors today are in East Africa, although there are also some very promising sites in Pakistan. Africa today is being torn apart by the forces of continental drift. This has been going on for the past 10 million years or so, and the great African rift valley marks the site of the spreading and so-called geotectonic activity. One day, if the process continues, the sea will rush in and flood the rift valley, which in 50 or 100 million years will form an ocean basin to rival the South Atlantic. That, fortunately, is in the far future. In the past this same geotectonic activity has been responsible for two vital parts of our story. First, it has produced great volcanic outbursts as molten material from the Earth's interior is squeezed to the surface by the geotectonic forces. Secondly, the deep valleys of the rift, between the volcanic mountains, have over the past several million years provided an ideal home for all sorts of animals, warmed by the tropical Sun and watered by the streams and rivers running off the volcanic mountains. Like many other animals, our ancestors found these rift bottom sites happy valleys rich in food and water and with a pleasant climate. Settling at the edges of the lakes or alongside the streams they inevitably left their remains in the silty muds of the water margins. Over millions of years a few of those remains were preserved as fossils, sandwiched in the rocks between layers of ash from the active volcanoes. All that was needed to complete the process was for a change in climate and the continuing geological changes wrought by the widening of the rift to isolate a few pockets of soft fossil-bearing rocks where they can be eroded today, and the fossil-hunters would be assured of their own happy hunting grounds.

At the Omo valley in Ethiopia, at Lake Turkana in northern Kenya, at Olduvai and Laetoli in Tanzania, the conditions we have outlined are met, and that is where hominid fossils are to be found, though not in great quantities. At Olduvai, made famous by Louis and Mary Leakey, there is a 300-foot-thick sandwich made up of layers of sediment and ash, formed over a period of about 2 million

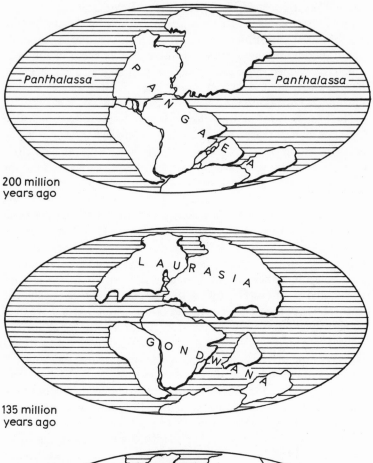

200 million
years ago

135 million
years ago

Present

In 50 million years' time

The evolution of life on Earth has been strongly influenced by the drift of the continents. The final break-up of Laurasia and Gondwanaland must have played a part in the events that led to the death of the dinosaurs and the rise of the mammals.

years. The whole sandwich can be read as a calendar of one of the most important periods in prehuman history, and the reading is made considerably easier by the river that flows through the gorge. The river created the gorge, slicing its way through 300 feet of rock, and as it did so it exposed sediments of all ages to view: though the actual task is a great deal harder, in principle one can examine 2 million years of antiquity by clambering down the gorge from the plateau to the river below.

Is it just luck that East Africa, the apparent birth place of the human species, should be one of the few places in the world where the combination of geological and climatic features preserved fossils in the past, provided a means of dating them, and now exposes them to present-day scrutiny? It isn't possible to say with any certainty, but it is surely a reasonable speculation that the geological activity in the region played an important part in the emergence of the man-ape line from the earlier monkey stock. Certainly something happened in Africa that did not in South America, and we know of no better candidate than a geological process that changed the environment dramatically and provided new opportunities among the rift valleys, tempting some of the early monkey primates to come out of the trees and seek a new way of life on the ground.

This move down to the ground was one of the crucial developments in the evolution of man and the apes. Under almost all circumstances animals tend to do better by being bigger, and with the dinosaurs out of the way the continued evolution of the prosimians shows this tendency at work. Small squirrel-like early primates could scurry about in the treetops and leap from branch to branch with ease, just as smaller monkeys still do today. But the larger monkeys could not move so freely in the trees, needing all four limbs to run along branches and using their tails to balance. Apes, bigger still, don't really leap about the trees at all; they spend some time, especially the gibbons and orang-utan, climbing through the trees, swinging by the arms, and some time walking on the ground, usually supporting themselves on the knuckled-under fingers of their hands. The two modes of locomotion — arm-swinging and knuckle-walking — are related, for the same anatomical adaptations that make for more efficient brachiation also enable the apes to walk both on all fours and on their hind legs with considerably more facility than the normal four-footed monkeys. Indeed, we've seen how arm-swinging is a fundamental ability of apes, all of whom can do it while no monkey can, and that (with the exception of man) the apes are present only in the Old World. The inescapable conclusion is that in the dense tropical forests of South America there was neither the pressure nor the opportunity for the basic monkey stock to change its way of life; in the Old World the ability to brachiate, and to move across open ground more easily, must have been an advantage.

Despite its membership of the ape family, *Aegyptopithecus* must have led a life more like that of a modern monkey than that of a modern ape, because the slender evidence of its sparse fossil remains suggests a body suited to running along branches, not swinging below them. While this raises the question of just when the apes did acquire their arm-swinging adaptations, that debate is not of the greatest importance here and now, for the very fact that an identifiable ancestor of the human-ape line lived in Africa some 28 million years ago is all we really need glean from the discovery of *Aegyptopithecus*. After all, the gap from *Aegyptopithecus* until the next oldest of our ancestors is a full 8 million years, so that as far as present-day palaeontology is concerned *Aegyptopithecus,* important though it undoubtedly is, sits in splendid isolation, a marker of the fact that by 28 million years ago the line that was eventually to lead to man and the apes had already become distinct from the monkeys. It is the first character in our story.

The problem of unravelling human origins is not made any easier by the fact that we have to accept the fossil discoveries that come our way as they are made, rather than as we would like them to come into the story. *Aegyptopithecus,* for example, wasn't discovered until 1960, but the next character in the story had been discovered 12 years earlier in 1948 by Louis and Mary Leakey on Rusinga Island in Lake Victoria. This was *Dryopithecus* (woodland ape), and its age was established at 20 million years, 8 million years younger than *Aegyptopithecus.* (Some fossils found in the 19th century are now classified as *Dryopithecus,* although at the time they were a great mystery.) The early discovery of *Dryopithecus* highlights the problem of taking the fossils as they come, but in our retelling of the story of human origins we shall, as far as possible, skim over the saga of the fossil hunters and even ignore the history of the fossils and try to set out the story as it is generally told, from *Aegyptopithecus* to the present day.

The name *Dryopithecus* is used to cover a fairly large and seemingly widespread family of ape-related ancestors that lived around 20 million years ago. Another genus, *Pliopithecus,* lived at the same time, and its remains have been found both in Europe and in Asia. Pliopithecines, as they are called, are generally thought of as ancestors of the modern gibbons, although the evidence for this is slender. They were 'gibbon-like above the neck, but were very different in their postcranial anatomy,' says David Pilbeam, eminent Yale palaeontologist. The major difference was that the arms and legs of the pliopithecines were roughly the same length, 'which,' Pilbeam says, 'implies that they were quadrupeds not brachiators'.[4] Nevertheless, because of the similarities in their skulls, the pliopithecines are normally accepted as the ancestors of the gibbons and siamangs. Certainly they are not ancestors of man.

What then of the dryopithecines? There are at least three species, probably derived from *Aegyptopithecus,* and the remains are judged to be pretty primitive. Because there is a better candidate for our own ancestor already alive at this time, the dryopithecines are normally given the task of representing the ancestors of the modern apes – excluding man. Pilbeam belives that one species, *Dryopithecus major,* is the ancestor of the gorilla, while another, *D. africanus,* could be the chimpanzees' ancestor. As to the orang-utan, Pilbeam admits that 'the fossil history of the orang-utan is obscure,' though he feels that one of the Eurasian dryopithecines 'might possibly have been related in some way'.[5] This part of the story, then, takes care of the emergence of the apes. They evolved from forms that were distinct some 20 million years ago, with the orang-utan splitting off even earlier. The naked ape is also presumed to have split off at a similar sort of time.

Unfortunately, there is no evidence that we know of that points conclusively to *Dryopithecus* as an ancestor of the modern apes. It may be, as always, that the fossils have not yet been found, but it could equally well be true that, in focussing on the human line of evolution, and eager to push that back as far as they can, palaeontologists have taken the easy way out with *Dryopithecus*, assigning him to a place where he does not belong. In any case, there is another contemporary of *Dryopithecus* called *Gigantopithecus*, whose fossil remains can be traced in rocks right up to half a million years ago, when it seems to have died out in the most recent great ice age. *Gigantopithecus* could easily have been the ancestor of the chimp and gorilla, but it is normally assumed that it wasn't.

This chapter, however, is supposed to be dedicated to the traditional story, and it is not too charitable to raise all our objections now; we hope they will all become clear before too long. To get back to the traditional interpretation of the fossils, we now come to the interesting bit, the beginning of our own ancestral line. It starts with a creature called *Ramapithecus*, found first in India and named after a prince in Indian mythology. *Ramapithecus* is known to us as a handful of jaw scraps and teeth and a bit of skull; there is none of his body skeleton, though in keeping with his status as man's ancestor he is usually drawn almost upright. The oldest ramapithecine fossils are about 14 million years old, and the conventional wisdom has it that some time during the long gap between *Aegyptopithecus* and *Dryopithecus* there lived a common ancestor of *Dryopithecus* and *Ramapithecus*. This missing link, probably around 25 million years old, would be the youngest common ancestor of man and the African apes, for by the time we find *Dryopithecus* in the fossils, according to the traditional picture, the line to *Ramapithecus* and man has already split off and become distinct. That all the ramapithecine fossils are younger than the dryopithecine fossils is just the luck of the draw; some day, the palaeontologists hope, a very old *Ramapithecus* will turn up.

How can palaeontologists tell, from the few bits of jaw and teeth, that *Ramapithecus* was a hominid rather than an ape? Their decision is based, naturally enough, on the shape of the jaw and the wear patterns of the teeth. *Ramapithecus* has teeth and jaw, and by implication eating habits, that are more like those of a man than those of a chimpanzee. Actually, of the three sets of jaws — man, chimp and *Ramapithecus* — man and chimp are the most different, with *Ramapithecus* somewhere in between. The closer similarity to the human jaw, of the other two, is *Ramapithecus*; the closer similarity to the chimp, of the other two, is also *Ramapithecus*. That, essentially, is how palaeontologists have arrived at the conclusion that *Ramapithecus* deserves the status of the first representative of the human family, the first hominid.

More cautious palaeontologists prefer to reserve the title of 'first hominid' for the creature that succeeds *Ramapithecus* in the fossil record. The gap between *Ramapithecus* and the next stage is about 6 million years, and it is only after that gap that it becomes possible to provide more than the sketchiest outline to the picture of human evolution. And the gap after *Ramapithecus* is much more frustrating than the gap after *Aegyptopithecus,* as Richard Leakey and Roger Lewin explain in their book *Origins*:

> 'On one side there is just one creature, *Ramapithecus,* while milling about on the other side is a menagerie of hominids, *Australopithecus africanus, Australopithecus boisei,* early *Homo,* and late *Ramapithecus*. And if we were to leap forward in time to perhaps three-quarters of a million years ago we find that the menagerie has dwindled once again to a single representative, a creature called *Homo erectus*.'[6]

This period from 4 million years ago to just under one million years ago is, then, at the heart of the story of the emergence of modern man. Unravelling it is made all the more difficult by the frustrating fact that there seem to be no clues about why or how *Ramapithecus* gave rise to the diverse hominid menagerie during the silent millions of years. And, according to the conventional view, whatever it was that set man off down a road different from that travelled by the apes took place not in the frustrating *Ramapithecus* gap but way back in the mists of pre-history, in the gap between *Aegyptopithecus* and *Dryopithecus* more than 20 million years ago.

When the first attempts were made to interpret the story of man's evolution from the fragmentary fossil record, it was thought that there had been a single line of hominid descent for the past several million years. Every hominid identified was regarded as fitting in somewhere as a direct ancestor of man, and palaeontologists were seeking, in the popular phrase, the missing link (or links) between man and ape. Today, with rather more evidence available, the experts see things rather differently. After the great diversification from the ramapithecine stock, only one branch of the hominid family leads to man. The others are dead ends. Some of the evolutionary stages along the direct path have been identified, in the form of *Homo habilis*, dwelling in East Africa around 3 million years ago, and *Homo erectus*, an almost completely modern man whose remains have been found all over the Old World in strata that range from about one and a half to half a million years old. *Erectus* is the descendant of *habilis*, and the key feature of this story is the existence so long ago of *habilis*, a creature showing many human characteristics which place it unmistakably on the trail from *Ramapithecus* to *Homo sapiens*.

Before fossil discoveries had shown the presence of *Homo habilis* 3 million years ago (and the key discovery, of an astoundingly complete skull that goes by the name of 1470, was made by Richard Leakey only in 1972) the only definite identifications of hominids from that period involved two species of another genus. These were *Australopithecus africanus* and *Australopithecus boisei*. (*Australopithecus* means southern ape, and the first one was the Taung child discovered in South Africa by Raymond Dart in 1925; the name might be a reference by Dart to his own homeland of Australia, as the species name of *africanus* commemorates its location as much as anything else. *Australopithecus boisei* − the original Nutcracker man − was discovered by Louis Leakey in 1959, and named for Charles Boise, an American-born businessman in London who contributed generously to Leakey's projects.) Both species of *Australopithecus* were more ape-like than *Homo habilis*, and until the discovery of 1470 either of them seemed a good candidate for the missing link. But if *Homo habilis* lived alongside *Australopithecus* − both figuratively, in that their remains are often found in the same geological strata, and literally, as they are also found at the same sites − then *Australopithecus* cannot have been the ancestor of *Homo habilis*.

The most problematic of these early hominid fossils are those found in Ethiopia by Don Johanson's team. They have uncovered a treasure trove of fossils that, after much thought, they classified as an entirely new species, *Australopithecus afarensis*. The famous Lucy is a member of this species, and she is tiny. The much larger bones found at the same site, Johanson says, belonged to the males. Richard Leakey disagrees with Johanson's interpretation, and each can muster impressive points in his favour, but at present we think it is impossible to decide. Another field season in Ethiopia, closed for years because of civil war, might settle the issue. For now, we can record that Johanson sees Lucy as the common ancestor of the other two australopithecines and the *Homo* lineage. Leakey prefers to trace our line back to an ancestor that gave rise to all the australopithecines and early *Homo*.

Homo habilis was an upright hominid 4 to 5 feet tall with a large skull, and he probably shared parts of Africa with *A. boisei*, who was about the same height but of thicker build and with powerful jaws set in a less rounded skull, and *A. africanus*, similar to *A. boisei* but without the nutcracker jaws and more slender, perhaps no more than 4 feet tall. There is just a hint that a fourth hominid may also have been around at this time, the last of the ramapithecines hanging on alongside the cousins it had given rise to. Neither *Australopithecus boisei* nor *Australopithecus africanus*, then, can possibly be the missing link, and the story of human evolution from 3 million years ago to the present day concerns the *Homo* line, with *Homo habilis* at its head. The two australopithecines died out, along with the last fading remnants of *Ramapithecus*, and *Homo* was left.

One of the most dramatic, and surely significant, discoveries about the hominid fossils after *Ramapithecus* concerns their location. Whereas *Ramapithecus* has been found in India, Africa, and even parts of Europe, where the climate 15–8 million years ago was a lot warmer than it is today, *Homo habilis* and his two australopithecine cousins are found only in Africa. Whatever it was that stimulated the fragmentation of *Ramapithecus* and gave rise to *Homo* happened only in Africa. Of course it is possible that this 'discovery', like so many others, is merely the result of the way the fossil-bearing rocks are located. Perhaps the strata 1–4 million years old and bearing hominid fossils that are so conveniently visible in Africa are buried every-where else. Evidence of *Homo habilis* may lie in 3-million-year-old Indian strata deep beneath the fossil-hunters' feet, but in truth this seems unlikely, especially when taken in conjunction with the modern evidence that man is so closely related to the chimpanzee and gorilla, for these species are found only in Africa. It looks very much as if the evolutionary changes which led to man happened in one small part of the world — stimulated perhaps by the environmental changes linked with volcanic activity along the rift valley — and that only in the past million years or so has *Homo* spread out from his East African cradle to cover the globe.

The discovery of *Homo erectus* remains at Koobi Fora on the shores of Lake Turkana in Northern Kenya, in rocks about 1½ million years old, confirms this picture. This new species of *Homo* is even more man-like than *habilis,* and differs very little from the specimen that lived a million years later in China — the famous Peking man. In a million years it would have been very easy for these early men to have spread from Africa to China; even if each generation travelled only 25 miles, it would take less than 10,000 years to complete the trek from Nairobi to Peking. At roughly the same time as *Homo erectus* was living in China, a different form had emerged in East Africa. This was Solo man, and he probably represents the first member of our own species, *Homo sapiens.* Our species, then, seems to have appeared, not quite in its final form, within the last 500 thousand years. *Ramapithecus,* remember, was around for at least 10 million years, which gives you some idea of how little time *Homo sapiens* has been in existence; our ancestry, according to the conventional view, is long indeed, but we are still children. A nailfile would eradicate modern man.

Before we go on to examine how and why such evolutionary changes have occurred, and to question the conventional story that we have outlined, there is one last minor, but perhaps highly significant, diversion on the long road from *Ramapithecus* to *Homo sapiens* that we ought to trace. This was the short-lived evolutionary experiment which saw the *Homo sapiens* stock split about 100,000 years ago, roughly at the beginning of the most recent great ice age, only for one line, *Homo sapiens neanderthalensis,* Neanderthal man, to disappear again within a few tens of thousands of years. Because this was such a recent diversification, we have ample fossil remains to tell us what went on, and to be able to describe changes happening over tens of thousands of years, rather than the customary millions. But such minor evolutionary branchings must have been happening throughout the history of human evolution. If Neanderthal man could come into existence, flourish briefly, and then die out, all in the space of less than 100 thousand years, then surely variations of the *Ramapithecus* line, or versions of *Australopithecus* or early *Homo*, must have emerged briefly many times during the past 10 million years in response to changing environmental pressures. We see only the successful lines in the fossil record, species that were around long enough, and in sufficient numbers, for their remains to be selected by the fossilisation lottery and for a few of those remains to come up a second time, in the palaeontologists' lottery. We seldom see the more ephemeral experiments of evolution, the short-lived variations on a theme that we can call, with hindsight, failures.

When the first remains of Neanderthal man were found, at almost the same time as Darwin's *Origin* was published, they created, as we have described, an almighty row between the evolutionists and the fundamentalists. The fundamentalists, however, were fairly quickly found wanting, and the evolutionists set to with a will to glean what they could of man's ancestry from the Neanderthal remains. The remains seemed to provide a convincing example both of a missing link and of an evolutionary dead end. One of the first, fairly complete, Neanderthal skeletons suggested an individual with a pronounced stoop to his back and permanently bent knees, as well as the more familiar beetle-brows and jutting head. The picture this conjured up, of a shambling man-ape, has coloured the popular image of our Neanderthal brothers even to the present day — but it is almost totally wrong. By one of the unkindest quirks of fate, and ironically, considering Professor Mayer's pronouncements about rickets, that early specimen was the remains of an individual who had suffered severe osteo-arthritis, which had the same potential to cripple then as it does now. He was no more typical of his species than the Hunchback of Notre Dame, or the Elephant Man, would be of ours. The wealth of Neanderthal remains uncovered since the mid-nineteenth century now reveals the typical Neanderthal man as completely upright, admittedly with prominent brow-ridges and a thicker skull, but with a brain slightly larger than modern man's.

Even so respectable a modern authority as Richard Leakey, writing with Roger Lewin in their book *People of the Lake*, can be found describing the Neanderthal people as a 'failure'[7]; yet the best evidence is that this minor deviation from our ancestral line shows very dramatically the success of natural selection at work. In any normal population there are some individuals with better eyesight, some who run faster, some more skilled at hiding, some better suited to cold weather, some to hot, and so on. The great mass of a population is made up of individuals who are broadly similar with only a small minority having an unusual degree of variation from the norm. In the middle of its natural geographic range, and with unchanging climate and other environmental factors, the species changes very little, but the potential for change is always there, in the variety of different minorities who between them preserve different kinds of abilities, keeping their options open.

Out on the fringes of the species' range, conditions are likely to be harsher than in the centre, and that is also where the first winds of change in outside pressures are likely to be felt, if a new predator appears, say, or the climate changes. Suppose, for example, that the weather becomes colder and drier. When this happens the minorities among the population who need less water and can withstand the cold will have a significant advantage, while the normal mainstream individuals will suffer. In a few generations, as the less hardy individuals are killed off and fail to breed, a new normal pattern will be established in the population, and the species will have adapted to the change. Then, as the extent of the environmental change, whatever it is, shifts and affects the species across its entire range, the now well-adapted members out on the fringes will be able to sweep across the range and replace their now poorly-adapted brethren. The fossil-hunter, examining remains in the middle of the range, will see a sudden transition from one form to the next, and may not realise that it was a change out on the fringes of the range that first promoted the evolutionary adaptation and then allowed that adaptation to spread in from the fringes and conquer the whole range.

Something very like this happened to man about 100,000 years ago. A severe ice age set in across the northern hemisphere and, especially at high latitudes out on the fringes of *Homo*'s normal range, individuals with certain exceptional abilities were favoured. Probably these features included a hairier skin and more fat, but all we know for sure we know from relics — bones, tools and pictures — left behind by the adapting men. Perhaps the survival characteristics included a bigger brain that enabled more intelligent individuals to cope with the harsher and more demanding environment. Whatever the details, within 25,000 years the *Homo* line had responded to the new evolutionary problems posed by the ice age with Neanderthal man. Neanderthal people were certainly intelligent — they used tools, painted pictures in caves, built shelters, and even buried their dead with ritual, judging by the evidence of a flower-bedecked grave found in Iraq — and they were probably much better adapted to the cold than modern man. If the ice age had lasted a little longer, Neanderthal scientists might even now be writing books about the evolutionary failure of our line. But the ice age did not last. The North warmed up again and the Neanderthals no longer had their advantage. True to the evolutionary pattern, however, *Homo* had kept its options open. What was bad for *sapiens neanderthalensis* was good for *sapiens sapiens,* who spread out over the entire globe as the ice retreated.

Rather than being a dead end or an evolutionary mistake, then, Neanderthal man represents an efficient and rapid response to changing conditions, an insurance policy against permanent environmental changes. It must surely be considered a success, looking at evolution as a whole. And the Neanderthals probably didn't die out entirely, in the sense that they have no descendants on Earth today. The split in the *sapiens* line was in all likelihood much too minor and short-lived to have driven a solid wedge between *neanderthalensis* and *sapiens,* who in all probability were able to breed successfully together. To members of the two groups, 40,000 years ago or so, the differences might well have seemed no more significant than mere tribal differences, less than the differences between a modern eskimo and an African pygmy. The best answer to the question of what happened to the Neanderthals is that they were absorbed by the expansion of *Homo sapiens sapiens,* with their remnants joining the spreading bands of modern men and becoming, once again, a minority variation within the main *Homo* theme. We may all number Neanderthals among our ancestors, and the human species as a whole may well remember the adaptations it needed in the ice age. When the new ice age comes, as it will with certainty, the fact that we count Neanderthals among our ancestors, and carry Neanderthal genes in our cells, may be our salvation.

That, then, is the traditional timescale of man's evolution. From the primitive *Aegyptopithecus* 28 million years ago through the mystery years to the jaws and teeth of *Ramapithecus* some 12 million years later. Then, another tantalising gap until, 8 million years later, about 4 million years ago, we have a positive menagerie of hominids wandering the East African plains. *Australopithecus* dies out, for some reason, leaving *Homo habilis* to carry the torch through *erectus* and *sapiens neanderthalensis* to *sapiens sapiens.* We haven't, here, outlined the traditional reasons given for these changes, nor have we explored the many differences of opinion that even traditionalists are prone to. But we would stress again that this view of man does not answer some of the most fundamental questions of our origins. If, as the story would have it, our line was already separate from that of the apes some 25 million years ago, why are we so very similar, genetically and anatomically, to the apes? The fossils certainly don't tell us. It might be that we all evolved these special features as a sort of cosmic coincidence, but it is far more likely that the solution to the monkey puzzle lies in our molecules, and in the evidence that we shared a common ancestor not 25 million years ago in the depths of the Miocene but less than 5 million years ago. And just as the fossils don't really provide all that much support for the traditional time-scale, they do not disprove the molecular clock, as we shall see.

Four
MODERN TIMES:
THE MOLECULAR CLOCK

The further from the DNA, the further from the truth. This epigram certainly has an authentic ring about it, and we would unhesitatingly agree with the unknown genius who summed up our story so well, for the fact is that evolution proceeds through the natural selection of genes, so that it is to the genes that we must look if we want to know the path that has been followed. The problem is rather like that of the medieval scholar who wants to trace the movements of books written in the days before Gutenberg invented movable type. At that time, all books were, quite literally, *written* by patient scribes. Even the most perfect scribe, however, occasionally made copying mistakes, and today bibliophiles pay particular attention to books that contain these errors. Historians can trace the spread of ideas from the spread of errors. If a volume written in Paris in 1350 contains exactly the same errors as one written in Warsaw in 1340 the scholar can be reasonably certain that the Paris volume was indeed copied from the Warsaw volume, which must somehow have been transported from Poland to France. And that shows where new ideas – new books – originated and how their messages spread. Just so with DNA and genes; if two genes are identical in every respect then they are each copies of one original. They share a common ancestor. By looking at differences between genes in species alive today, biologists can get a very good notion of how nature's new 'ideas' – mutations – appeared and how species have diverged.

DNA is the ultimate ancestor, not only of every chemical that is manufactured by the living body but of all living bodies. There is no 'chicken and egg' problem here; DNA came first and the complexity of living cells and bodies followed. It is the sequence of nucleotides along the DNA that determines the sequence of amino acids along proteins which in turn determine how the body will carry out its functions, so the ultimate truth of molecular evolution is contained in the sequence of nucleotides along the DNA, the words of the genetic code. The sequence of amino acids along the protein is only one step removed from this basic code, and also tells us a great deal about the relationships between species. If two proteins have identical sequences the DNA that codes for them may still not be quite identical because of the redundancy of the genetic code. The fact that there is a leucine at the same point on two protein chains says nothing about the codons at the corresponding points on the DNA; they might be the same, but they could easily be different and we have no way of discovering, from the amino acid sequence alone, the nucleotide sequence on the DNA. One step further removed still are the immune properties of the protein, which make up its pattern of behaviour. Just as the amino acid sequence is a good guide to the underlying nucleotide sequence, so too the immune response is a good guide to the underlying amino acid sequence.

Those, then, are the levels of analysis at which the molecular evolutionist works. He can focus on the ultimate truth contained in the sequence of bases along the DNA, move out a little further to the amino acid sequence of the protein commanded by the DNA, or move still further out to the response of a foreign immune system to that protein. It is easier to work from the outside inwards, and that indeed is how the study of molecular evolution developed historically. But we emphasise that this does not mean that the two protein-centred levels are any less informative than the DNA itself, just that, like everything else about evolutionary biology, the further from the DNA, the further from the truth.

The concept of molecular phylogeny – evolutionary history according to the molecules – is simple. Measure the differences between a number of animals, preferably at the DNA level, and you are ready to construct an evolutionary tree that will tell you how the animals in question are related, which of them share recent common ancestors, and how long ago they split from the common stock. In practice, of course, things are not so simple, and in this chapter we will look at the various methods of the molecular wizards as well as at some of their more notable successes. We'll begin at the beginning, with an update of the method that started the whole ball rolling, evolutionary immunology.

George Nuttall, whom we met in Chapter One, obtained excellent results 80 years ago by injecting rabbits with whole serum to produce antibodies and then mixing this with a variety of sera to see how strongly they reacted. Vincent Sarich, in his laboratory above the Berkeley campus, has brought the use of immunology in the study of evolution to a fine art, although the story remains in essence the same; create an antibody to one antigen and measure the strength of reaction with various related antigens. Looking out over the leafy campus and San Francisco Bay from his fourth floor laboratory in the squat, redbrick biochemistry building, Sarich seems a long way from the grimy Cambridge laboratory that housed Nuttall, but he is Nuttall's spiritual heir. Always on the lookout for samples from new and interesting species that he can slot into his ever-growing store of antigens and antisera, Sarich continually buttonholes colleagues for bits of their unwanted animals. 'Blood will do,' he tells them, 'but I'd rather have meat if at all possible.'[1] When he gets the sample, blood or meat, it begins a long journey through a complicated series of procedures designed to attract and purify the proteins of interest and then elicit very powerful antibodies to those proteins. The first stage is to isolate the proteins, usually albumin and transferrin. These are big molecules, and preparing them is not too difficult. The blood serum is first diluted and then heated, which tends to make the proteins coagulate, just as egg-white (which is almost pure albumin) coagulates when you boil an egg. Extra purity is achieved by so-called chromatographic separation, in which the mixture of partially purified proteins is poured into a special column of carboxymethyl cellulose, really just a superior form of filter paper. If Sarich takes care to add just the right amount of other chemicals then most of the other proteins stick to the column but the albumin moves through it, like the wine stain on the table cloth, and from a mixture of proteins at one end, pure albumin emerges at the other.

Pure albumin is a good beginning, but it is only a beginning. Now comes the tedium and routine of using some of that albumin to prepare antibodies. As in Nuttall's day, the animal of choice here is the rabbit. Sarich uses male white rabbits of the New Zealand strain as these have a vigorous immune system that makes lots of antibody, and years of experience have allowed him to evolve a timetable that produces the finest quality antiserum from tiny amounts — a few thousandths of a gram — of antigen. It takes four rabbits and nearly four months, with injections of albumin antigen after 5, 7, 12 and 13 weeks. After this a small amount of each rabbit's blood is removed and the antibodies in it tested and then purified in much the same way as the albumin was purified. That done, the antisera are kept until needed. Sarich's Berkeley laboratory is packed with freezers, each filled with neatly labelled bottles of purified antisera from hundreds of species, exotic and commonplace. There is the binturong and the dog, giant panda and grizzly bear, mongoose and tabby cat, all stored at 10 degrees Celsius below freezing.

Meat and blood are Sarich's raw materials; pure antigens and pure antisera are the refined products with which he can really go to work. Nuttall estimated the strength of the reaction from the cloudiness of his solutions. Morris Goodman in Detroit used the immunodiffusion technique, which enabled him to rank antigens into a linear order of similarity. Sarich wants numbers, meaningful quantitative measurements of the antigenic similarity of two different albumins that go beyond subjective judgements, and he gets them.

Sarich's measuring system — called the micro-complement fixation test — is accurate but complex, and not so simple for non-experts to understand. Understanding, fortunately, is not important for non-experts. The essence of the technique is that it takes advantage of a cascade of naturally occurring magnifying effects to increase the sensitivity of normal immunological techniques thousands of times, and this means that tiny quantities are sufficient. 'It's not just that it's more sensitive than the existing techniques,' Allan Wilson told us, 'though the fact that it could detect a single amino acid substitution was pretty astonishing and it was the thing that led us to use it. It was extremely economical of materials. Nuttall and other people would run out of the serum that they had,' but with micro-complement techniques Sarich and Wilson had 'a reagent that would last for decades.' Wilson explained that they could 'look at hundreds of species with one antiserum'. Best of all, 'if more species came along later, you could go back to the freezer and get it out, and it would still be unaltered'.[2] The technique is a masterpiece of practical biochemistry, at the end of which Sarich has the numbers he has been seeking, a measure that he calls the immunological distance, which relates the extent to which proteins from two different species differ from one another. The immunological distance (often referred to as ID) ranges from 0, for identical species, to about 180, for the maximum unrelatedness measurable by the technique of micro-complement fixation. Man and the domestic cat, for example, are about 173 units apart on this scale.

So, after all the extraction, the purification, the mixing and the

measuring, Sarich and Wilson obtain a single number, the immunological distance, that represents the relatedness of two proteins, and
hence of the two animals that provided the proteins. The
immunological approach is necessarily indirect, but it nevertheless
gives results that are broadly in accord with palaeontological theory.
In most cases, where the palaeontologists believe two species are
more closely related to one another than to a third, the immune
approach agrees. By and large, the technique refines the picture
palaeontologists already have by making more numbers available.
As we have seen, however, it does produce some surprises, not the
least of which has been the revelation that man, chimpanzee and
gorilla are all equally close. But the immunological approach is not
the last word. It is removed from the DNA by several steps, and
there are some molecular evolutionists who prefer not to be those
few steps away from the DNA. They look to the protein itself, and
in particular to the actual sequence of amino acids.

The ability to determine the sequence of the hundreds of amino acids
that make up the average protein is one of the great, and largely
unsung, success stories of science. The chemistry involved is
astounding. You not only have to identify each amino acid present,
but also decide just where in the chain each one goes. In the early
days of molecular biology, 40 years ago, people weren't even sure
that an individual protein had a fixed structure. When Fred Sanger,
now at the Laboratory of Molecular Biology in Cambridge (at the
time it was part of the Cavendish Laboratory) set out in the late 1940s
to determine the sequence of amino acids in a protein it seemed to
many that he was on a wild goose chase. Years later Sanger caught
his goose, the complete sequence of the hormone insulin, and an
additional trophy in the shape of a Nobel Prize. (Twenty years later
he pulled off the same *coup* again, and achieved a second Nobel Prize
for sequencing nucleic acids, which again everyone had thought was
impossible.) Getting the sequence of insulin initially took a huge
effort, and a vast number of man-hours; today it is relatively easy to
sequence a protein, and the whole job can be completed routinely in
a matter of days rather than years. In general, though, molecular
evolutionists tend not to sequence their own proteins; there are
plenty of molecular biologists around who are frantically sequencing
proteins for other reasons entirely, and of course they publish their
results in scientific journals, so there is a great deal of data available to

anyone who wants to study the evolutionary history of proteins through their amino acid sequences. The first step towards such sequences came from studies of haemoglobin, studies aimed at answering the mystery of sickle-cell anaemia rather than at any insight into evolution. The technique, called fingerprinting because of the unique way it identifies proteins, was developed in 1955 by Vernon Ingram, like Sanger working at the Cavendish Laboratory in Cambridge.

Fingerprinting provided one of the first hints that proteins did indeed have a fixed and unchanging structure and that changes in the DNA were related to changes in that protein structure. At that time, chemists knew that normal and sickle-cell haemoglobins moved at different speeds through an electric field, showing that there must indeed be some physical difference between them, but no chemical tests on the whole molecules were sensitive enough to show up the difference, whatever it was. Ingram decided to break the haemo-globin up into smaller fragments, which might be more manage-able, and see whether there was any difference to be detected in the fragments.

We now know that haemoglobin is made of two different protein chains, one 141 amino acids long and the other 146, and that the chains are twisted and joined together. The first step in separating the molecule into smaller pieces is to keep the haemoglobin close to boiling point for several minutes, which destroys the folded struc-ture of the molecule and opens the chains up. Then Ingram added a small amount of another protein, trypsin, and kept the samples at body temperature for about two days. Trypsin is one of the digestive enzymes; it breaks apart the peptide bonds that hold amino acids together in a protein chain, but it does so selectively. Instead of breaking the peptide bonds at any old place in the chain, trypsin attacks only the peptide bond next to two specific amino acids, lysine and arginine, and it always chops on one side of the lysine or arginine, never on the other. There are about 30 such sites on the single haemoglobin chain so that a trypsin digest of haemoglobin always ends up as a mixture of about 30 sorts of pieces, and always the same 30. The average piece is about 10 amino acids long, and, as Ingram explains, 'the point is that a small difference between two complete proteins becomes a relatively large difference in a short segment, and that's the crux'.[3]

To look for the odd man out, the fragment that must differ

between sickle-cell and normal haemoglobin, Ingram used a combination of chromatography and electrophoresis. In chromatography the suspect chemicals are allowed to move through a filter paper of their own accord; in electrophoresis the chemicals are forced through the paper by an electric current, positively charged fragments attracted towards the negative pole and negatively charged ones to the positive pole. In both cases the rate at which a fragment moves depends on its composition — small, highly charged fragments moving faster than larger, more neutral ones. Ingram now took two pieces of ordinary filter paper; onto each he put a drop of trypsin digest, one piece getting the normal haemoglobin and the other the fragments of sickle-cell haemoglobin. He clamped each paper between two glass plates and put the plates into a bath of acidic solvents. Then he attached electrodes to the paper on either side of the spot and passed a current between the electrodes for some two and a half hours. The current separated the fragments, pulling most towards the negative electrode but leaving the neutral fragments in place and pulling two negatively charged fragments to the positive electrode. The fragments were partially separated, but there was still a great deal of overlap between them; to get rid of the overlap, Ingram turned the paper through 90 degrees and put the fragments through a chromatographic separation. He pinned the paper so that the line of fragments trailed in a pungent mixture of solvents and allowed the solvents to climb up the paper. As they did so, they carried the fragments with them, the smallest fragments being carried fastest, just behind the advancing line of the solvent. When the solvent reached the top of the paper, the separation was complete. Ingram could unpin the paper and set it aside to dry.

At this point, all Ingram had to show for his efforts were two bits of filter paper, white except for a few pencil scribblings. Like the detective who must brush powder on the doorknob to capture invisible fingerprints, he had to make the protein fragments visible. This he did with a chemical that marked the protein fragments with an orange stain. And when he stained the papers, he found 30 blobs, arranged in a rough triangle. Up the middle of the triangle was a backbone of blobs, some still slightly overlapping. These were the neutral fragments that had not been moved by electrophoresis and had simply spread out in a column during the chromatographic separation. To the right of these were two smaller blobs, representing the only negatively charged fragments, and to the left of the backbone were 18 other irregular blobs, fragments of different sizes but all with a positive charge that had been separated both by the electrophoresis and by the chromatography. Or rather, there were 18 fragments in the fingerprint of normal haemoglobin; the digest of sickle-cell haemoglobin contained the same total number of fragments — 30 — but the fingerprint differed in one, and only one, respect. To the left of the backbone, just a little way towards the positive electrode, was an additional spot. There were 19 negatively charged fragments, and a close look showed that the extra negative spot had been displaced from the neutral backbone. The difference had been found, and had been confined to just one fragment. With that fragment identified, further analysis soon showed that the negatively charged amino acid valine had been substituted for the neutral glutamine at the sixth position on one of the chain of normal haemoglobin. Just one letter had been changed in one word of the genetic code, and that word had been identified. The fingerprint technique that Ingram had developed could be used to identify proteins with the same degree of precision as real fingerprints could be used to identify people. A protein's fingerprint never changes, just as a person's are fixed from even before birth. And different proteins have different fingerprints, just as different people do.

Ingram, of course, was not particularly interested in using his technique to compare the proteins from different species, but he had demonstrated that the protein fingerprint was intimately linked to the DNA sequence. Others took up the technique specifically to study the evolution of protein molecules. One of these was Emile Zuckerkandl, the biochemist who first suggested that molecular evidence might be used as a clock. Indeed it is Zuckerkandl who is generally credited with inventing the phrase 'molecular anthropology' at a conference in 1960 at the lovely Austrian castle of Burg Wartenstein. The occasion was a meeting organised by the Wenner-Gren foundation to discuss Classification and Human Evolution, and it was one of the first academic gatherings at which the evidence of evolutionary history provided by molecules was aired. Zuckerkandl's contribution was called, simply, Perspectives in Molecular Anthropology, and he talked about the various ideas then current. More specifically, he presented his own evidence, which was based on a fingerprint analysis of the haemoglobins from a number of primate species. Using methods very similar to those of Ingram, Zuckerkandl and Linus Pauling had examined the haemoglobins of a variety of animal species. As he told the conference participants, 'gorilla, chimpanzee, and human hemoglobins were indistinguishable . . . whereas the pattern given by orang-utan hemoglobin differed slightly from the patterns characterizing man, chimpanzee and gorilla'. He also pointed out that the fingerprint pattern of rhesus monkey haemoglobin 'differed more significantly, although the similarities predominated largely'.[4] Zuckerkandl and Pauling didn't confine themselves to primates, they also looked at other species and concluded that in Artiodactyls (hoofed animals) there was a whole group of protein fragments that turned up in different places from those they occupied in human haemoglobin, but that there were still some notable similarities, whereas when they extended the comparison to groups as distantly related to us as the fishes, there was not a single fragment in the same place.

In the early 1960s sequencing a protein was still a time-consuming business, but naturally people decided that the haemoglobins of the apes were worth the effort, and Zuckerkandl told the Burg Wartenstein conferees that he and a colleague were in the process of determining the exact sequence of amino acids in the haemoglobin of the gorilla. Even then — 20 years ago — all the indications were that there was but a single difference in one of the chains between man and the gorilla, an aspartine in the gorilla at a point where there is a glutamine in man. A single amino acid difference, 'the smallest possible'. In other words, the differences between human and gorilla haemoglobins 'are of the same order — namely one single amino acid substitution — as the differences that have so far been found between humans with normal and abnormal hemoglobin chains. Therefore,' Zuckerkandl stressed, 'from the point of view of hemoglobin structure, it appears that gorilla is just an abnormal human, or man an abnormal gorilla, and the two species form actually one continuous population.' He went further, and speculated that 'it is possible that a gorilla hemoglobin chain exists as a mutation in some humans. The reverse — the existence of a human hemoglobin chain in some gorillas — is equally possible in principle, but less probable because of the comparatively small number of gorillas in existence.'[5]

The fingerprinting technique can provide a great deal of information about the construction of various proteins, but in the final analysis it is only about as useful as the immunodiffusion technique. It gives a qualitative feel for the distance between species, but doesn't actually put numbers to that distance. For a better source of information about the sequence of amino acids that make up a protein the molecular biochemist must turn to the amino acids themselves.

Several techniques have now been developed to identify the amino acids making up a protein chain. Most of these are variations on the use of digestive enzymes like trypsin to chop the whole protein into a set of smaller fragments that could be analysed by conventional methods. One digestive enzyme always chops the protein into the same set of fragments. Another one, equally specific, chops the chain into a different set of fragments. This set too can be analysed, providing the patient biochemist with two completely sequenced sets of fragments. The two sets of fragments, obviously, could each be joined up to make a complete protein, and the rest of the endeavour is devoted to playing jigsaw puzzles with the sequenced fragments, using the overlap between the sets to provide clues about the structure of the whole molecule. The first protein to be completely sequenced — to have every amino acid and its place on the chain identified — was the pancreatic hormone insulin, the work for which Fred Sanger won his first Nobel Prize. Since 1955, when Sanger published the insulin sequence, hundreds more proteins have been broken down in the same way, providing a welcome windfall for the molecular evolutionist.

Two pioneers in the use of amino acid sequences to trace evolutionary history are Walter M. Fitch and Emanuel Margoliash of Northwestern University near Chicago. The first protein they worked on was cytochrome c, one of a family of enzymes that all living things use to provide themselves with the energy they need. Biochemists investigating the cytochromes worked out the three-dimensional structure of the molecule and its amino acid sequence, first in the horse and then in a couple of fishes, the tuna and the bonito. By the time Fitch and Margoliash began their study, in the mid 1960s, they could call on data from more than 20 different species, and today there are scores more. The species they looked at included insects, fish, reptiles and mammals, and of course Fitch and Margoliash were aware of the accepted evolutionary history of the species whose cytochromes they were examining. Poring over the long computer print-outs of the various cytochrome sequences, they lined the various sequences up and looked at the specific amino acid in each position. Where the protein sub-units differed, as they did at

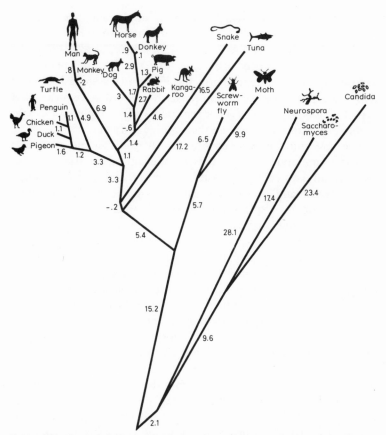

One of the first evolutionary trees based on molecular evidence. Compiled by Walter Fitch and Emanuel Margoliash from the amino acid sequences of the enzyme cytochrome c. This tree shows the order in which species diverged but does not provide dates — the numbers represent the minimum number of changes to the genetic code to produce the observed differences. (Modified from *Scientific American*, September 1978, p 69.)

a number of sites, Fitch and Margoliash worked out the DNA codons for that site and then calculated all possible ancestral codons that could give rise to those two codons with the minimum number of changes. For example, if there was a methionine at one site, the corresponding codon had to be TAC. At the same site in another species, the amino acid was valine; the codon for this could be CAA, CAG, CAT or CAC. The common ancestor of these two codons might be either TAC, which requires a single mutation to get to CAC, or it might be CAC, which by the same token requires but a single change to get to TAC. It could, of course, be something like GAC, which would need one change to get to TAC and another to get to CAC, but as a general principle Fitch and Margoliash adopted what they called the maximum parsimony principle, which stated that they would always favour the codon which required the least number of changes to produce the required end states.

The maximum parsimony principle is really an extension of a principle first enunciated in 1324 by William of Ockham, and for that reason known as Ockham's razor (because it cuts through philosophical tangles). Ockham's razor proposes simply that whenever one tries to explain some event or other one keeps to the minimum number of assumptions required. If two or more explanations both fit the evidence, then we choose the simplest. For example, if one can explain the movement of the planets through the skies by assuming that they travel in relatively simple paths around the Sun, it is preferable to do this than to assume much more complicated paths around the Earth. The problem facing the molecular evolutionist is to identify correctly the unobtainable ancestral amino acid sequences (or, more properly, the DNA sequences) given only the sequences in living species and making as few assumptions as possible. Then he has to derive a family tree showing how present-day forms evolved from the ancestral line. Actually, there are two separate problems. One, called the little maximum parsimony problem, is to construct the sequence of bases along the ancestral DNA given not only the amino acid sequences but also knowing in advance the shape of the tree. This isn't too difficult, and one of Fitch's great contributions was that he devised a computer algorithm that would indeed calculate the necessary DNA sequences of every ancestor in the tree, and do so correctly, as was subsequently proved mathematically. But the big maximum parsimony problem is much more tricky. This is to construct not only the ancestral DNA sequences, but also the shape of the tree itself, given only the amino acid sequences of the living branch tips of the tree.

The big maximum parsimony problem is one of an interesting family of mathematical problems that are all related and that are all very very difficult to solve. A travelling salesman, for example, who wants to calculate the shortest route that will take him to each of the towns on his list, faces a mathematically identical problem to Fitch when he wants to find the tree and the DNA sequences, and at present the only solution is the mathematical equivalent of sheer brute force. You have to try every possible solution and then look for the shortest one (there will always be only one). Now for a small-time salesman, or a tree with few branches, this approach can be put to work. For example, a tree with five living branches has only 15 possible arrangements, and one with six branches has 105 possible solutions. The drawback to the brute-force approach is that as the number of branches increases, the number of solutions explodes almost unimaginably. There are 945 different trees with seven branches, and 10,395 with eight branches. By the time you reach a tree with 20 branches the number of solutions, every last one of which must be looked at, increases to more than 10 to the 20th power (100,000,000,000,000,000,000). The fastest computer, which could evaluate one solution every millisecond, 60 thousand solutions a minute, would still take 10 thousand million years — roughly the age of the Universe — to find the most parsimonious tree for just 20 species, working backwards from the present-day family relationships alone.

There are, of course, ways around the problem, and mathematicians have devised methods that will provide a solution that is probably the best one in any real case, but at the moment there is no proof that the solution found by one of the short-cuts is truly the most parsimonious one for that set of data. In fact G. William Moore, one-time colleague of Morris Goodman and the man who proved that Fitch's solution to the little maximum parsimony problem was correct, has gone so far as to assert that 'it is safe to say that a definitive solution to the big maximum parisimony problem is not forthcoming in the near future'.[6] In spite of all the problems, though, Fitch and Margoliash did attempt the near impossible. From their scrutiny of the amino acid sequences of their 20 species, they constructed a tree that linked the 20 species to a single common ancestor and showed how closely any two species were related. The tree cannot be proved mathematically to be the best possible reconstruction, but it can be proved to be a very good approximation to the truth. It is remarkably similar to the conventional tree constructed on the basis of the anatomy of the species, and that is as it should be. There were a few surprises and disagreements of course, with chickens apparently more closely related to penguins than to pigeons or doves, and the turtle, a reptile, apparently closer to the birds than to the rattlesnake, another reptile. Some of these anomalies are just that, anomalies, but others, like the position of the turtle, have gone a long way towards strengthening the argument of a number of conventional — that is to say anatomically biased — evolutionists who say that the group we call reptiles is not in fact a meaningful group at all, and that the turtle is indeed closer to the birds than to the other reptiles that we call snakes. There are other methods, too, for constructing a tree from raw amino acid data, but in general the results they produce are all very similar, which suggests that there is indeed a great deal of truth in the molecules and that the various methods all approach that truth more or less accurately.

However, the problem remains that even the amino acid sequence is not the fundamental material of evolution; we are still a small distance away from the DNA itself. No matter how big or fast the computers, how sophisticated the mathematical algorithms, there remains the objection that we are nevertheless still guessing at the actual sequence of bases along the DNA. That is where the mutations that permit evolution take place, and that is where we should be looking if we want to see how evolution progresses. G. William Moore handles this problem with what he calls the Red King principle. The allusion is to Lewis Carroll's masterpiece *Alice in Wonderland*, a work cherished by most mathematicians. Alice, you will recall, ends up in court at the trial of the Knave of Hearts. 'Alice had never been in court before,' Carroll tells us, 'but she had read about them in books, and she was quite pleased to find that she knew the name of nearly everything there. "That is the judge," she said to herself, "because of his great wig".' The point is that Alice, who has never seen a judge before, has to infer that the man with the 'great wig' is a judge. He is, in fact, the King of Hearts, but because he *looks* like a judge, he must *be* a judge. As Moore explains, 'in examining . . . sequences from contemporary species (obtained by inference from amino acid sequences and the genetic code), we assume that when two sequences *look* alike they must *be* alike, in the sense of sharing a common ancestry'.[7] Granted the power of the Red King inference, it would still be nice to make that final step and come down from the amino acid sequence to the DNA itself, but just as the fingerprint technique is intermediate between immune response and amino acid sequence, so too there is a technique that is intermediate between amino acid sequence and nucleotide sequence. It is called DNA hybridisation.

The idea behind DNA hybridisation is simplicity itself. We separate the double helix of one species' DNA into its component single complementary strands. We do the same to the DNA of the species we wish to compare it with. Then we mix all the separated strands together. Where the two species have identical sequences along their DNA, the complementary bases from the two different species will be able to come together and join. In fact, if the two sets of DNA were completely identical, then we would end up with three different sorts of DNA after the mixing. Two sorts would be identical to the original, composed of complementary strands that had initially been in the same sample of DNA, though we would be very surprised if any particular strand managed to join with its original partner. Some might, of course, but in any case this doesn't matter as the double strands would be identical in every respect with the original sample, even if the pairs that come together in each new double helix are not original pairs. In addition to this reconstituted DNA we would also find in our sample an equal amount of DNA in which one strand had come from one sample and the other from the other, but unless we had previously labelled one of the lots of DNA we would not be able to tell which double-stranded molecule was which. If one sample had been made highly radioactive, and if we could count the radioactivity of each newly-formed double helix we would find that a quarter of the molecules were as radioactive as the original sample, and another quarter were not at all radioactive. These would be the molecules that had by lucky chance joined up with a complementary strand from their own sample. The remaining half would have half the radioactivity of the original sample, and these would be double-stranded molecules composed of one strand from the labelled sample and the complementary strand from the unlabelled sample.

That is how things would work if the two samples of DNA were in fact absolutely identicial. What would happen if this were not the case? Well, if the two samples were very similar, so that long stretches of each did indeed contain almost the same sequence of bases, then they would still be able to form so-called hybrid molecules. What we then want to know is the extent of the similarity, the percentage of the two DNAs that is in fact identical. To do this we make use of the fact that it is only the binding between individual pairs of complementary bases that holds the two strands together.

The more bonds there are, the harder it will be to separate the strands of the hybrid DNA, and to measure the strength of the bonding between the strands we need only put in energy to overcome the binding energy and see when the strands drift apart. What we are really doing is finding the melting point of the DNA. Solids are held together by bonds between their component atoms. Inject energy into the solid, in the form of heat, and you break the bonds so that the solid is free to become a liquid. The energy needed to break the molecular bonds is the melting point of the substance. To get back to DNA, if the two strands are totally complementary they will be held together with a certain strength and it will take a certain amount of energy, a temperature of around 85 degrees Celsius, to separate them. If the strands are only partially complementary the force holding them together will be weaker and it will take less energy, a lower temperature, to make them drift apart. Impure DNA, in the sense of being made of non-identical strands, has a lower melting point than pure DNA, just as impure water has a lower melting point (freezes at a lower temperature) than pure water.

Naturally, like all the techniques we have discussed, the theory of DNA hybridisation is easy enough to understand but putting it into practice is the devil's own job. Quite apart from the problems of extracting and purifying the DNA of the species we want to compare, there is the very tricky business of detecting the melting point of the hybrid. It isn't as simple as keeping one eye on a thermometer and the other on a block of ice, for although DNA can actually be seen under an electron microscope after suitable preparation it is impossible to tell when the strands in a solution of DNA have separated. Once the scientist has made his hybrid molecules, also known as heteroduplex DNA (which just means double-stranded DNA from different sources) he has to measure very accurately the temperature at which the double helices break down. The procedures depend heavily on the enormous advances that have been made in the technology of genetic engineering, a full discussion of which would be out of place here. We can, nevertheless, give some idea of how it is done.

The first step, as we've said, is to purify the DNA from the two species. Then, one species' DNA is labelled with a radioactive tag, usually the radioactive isotope of iodine; this does not affect the way the strands work, but simply acts as a tracer that enables researchers to follow the single strands from that species. With the

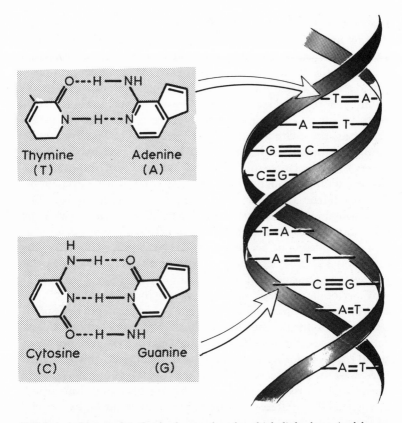

DNA is held together by hydrogen bonds which link the paired bases thymine-adenine and cytosine-guanine. The melting temperature of DNA is the temperature at which these bonds break and the strands of the double helix go their separate ways. If individual strands of DNA from different but closely related species are brought together, they form more loosely bound double helices because the base pairs are not perfectly matched. So the melting temperature is reduced in inverse proportion to the relatedness of the two species.

labelling done, the two sets of DNA are mixed and slowly heated. At around 85 degrees Celsius the bonds between opposite bases, which normally hold the strands together, are broken, and the strands drift apart. Now the mixture is allowed to cool slowly so that heteroduplex molecules can form from the two species of DNA. Once the mixture is cooled the few remaining single strands are removed and the business of measuring the melting point begins. The temperature is raised by about one degree and the dissociated DNA removed and assayed for radioactivity with an improved version of the old-fashioned geiger counter. Then the temperature is raised another notch and the next lot of single strands removed and counted. A repeated series of counts at steadily increasing temperatures produces a so-called dissociation curve, the peak of which represents the melting point of the hybrid DNA. This can then be compared directly with the melting point of a pure hybrid, that is DNA from the target species heated and allowed to recombine, so that any quirks due to the heating processes and so on are evened out. The size of the difference in melting points between heteroduplex and normal DNA is directly related to the dissimilarity between the two strands. A difference of one degree Celsius is roughly equivalent to a difference in one per cent of the DNA; one in a hundred of the nucleotides are not identical in two species that show a melting point depression of one degree Celsius.

Many biological relationships have now been explored with the hybridisation, or annealling, technique, which turns out to be particularly useful for illuminating the small differences between closely related species. We shall be discussing some of these in a little more detail later, but for the moment it is worth noting that two different species of mouse, placed by taxonomists in the same genus, have a heteroduplex dissociation temperature five degrees Celsius below normal, a DNA difference of about 5 per cent, while for two species of fruitfly the figure is much higher, 19 per cent or so. It is worth adding, too, that these pairs of species, which differ by perhaps a twentieth and a fifth of their DNA respectively, are extremely difficult to tell apart just by looking at them.

Heteroduplex DNA made from *Pan troglodytes* and *Homo sapiens* melts at a temperature just one degree Celsius below that of pure *Homo sapiens* DNA. Ninety-nine out of a hundred bases are identical in man and the chimp, which are not put in the same genus, or even the same family.

DNA annealling is a very powerful technique, but like protein fingerprinting it is a step away from the information we are really after. The ultimate truth resides not in the melting point of mixed DNA molecules but in the sequence of bases along the DNA itself. These are the dice that evolution rolls, and here we will find the successful moves that set one species off from another. Technology, again, is everything. If you've managed to extract the particular bit of DNA you are interested in from a variety of species, pure enough and in sufficient amounts to work with, actually uncovering the sequence of nucleotides is probably the easiest part of the whole business. That doesn't mean that it is simple, but such advances have been made in the genetic engineer's toolkit that one can think realistically of knowing the entire nucleic acid sequence of an organism. Indeed this has already been achieved, notably for the virus phiX174, whose complete DNA sequence occupied several pages of the scientific journal *Nature* in 1980. There are two methods now in common use for identifying DNA sequences, which won for their inventors a share in the Nobel Prize for 1980. Walter Gilbert, of Harvard University, got one for his method and Fred Sanger, at the Laboratory of Molecular Biology in Cambridge, got one for his. The 1980 prize put Sanger into the elite ranks of those few scientists who have earned the ultimate accolade not once, but twice; in fact Sanger is only the second person to have won the prize twice in the same field (the other is physicist John Bardeen of the University of Illinois at Urbana, who won in 1956 and again in 1972).

The entire DNA sequence of a virus is much shorter than a human genetic blueprint, and molecular biologists are still a long way from being able to write down the DNA sequence for a man, or any other mammal. But the technique is essentially a refinement of the molecule chopping and reconstruction techniques used to sequence proteins. DNA molecules are cut into pieces, labelled with radioactive tracer atoms, the fragments separated from one another by electrophoresis, and the sequence reconstructed. Of course the work is more difficult than protein sequencing, operating on a smaller molecular scale. It is rather as if we had described how to make a grandfather clock and then said that making a Cartier watch involved the same physics. It does, but our statement doesn't do justice to the watchmaker's craft, just as we cannot do justice here to the work of Wally Gilbert and Fred Sanger. For our purposes all that matters is that once the complete sequence of nucleotides along the DNA of different species is known, the evolutionary history of those species can be explored right down at the very level at which evolutionary changes take place.

No more worries about immune reactions, or silent substitutions, or potentially inaccurate melting points. The nucleotide sequence is the mother lode. Of course there are occasional problems, like finding great chunks of DNA inserted into a gene where you didn't expect them, or discovering that a simple sequence repeats itself almost endlessly along the molecule, but by and large it is relatively easy to use DNA sequence data to construct evolutionary trees. In one way it is easier to construct trees at this level than at the more distant ones, just because there are so many fewer assumptions to be made.

We outlined some of the pitfalls waiting for anyone attempting to reconstruct evolutionary trees from molecular information when we discussed the Red King hypothesis and the two maximum parsimony problems, but while it may seem a difficult job it is not an insuperable problem, especially with the help of a large high-speed computer to try out all the possible alternatives. Not surprisingly, then, there have been some notable successes with the molecular methods, which has been used to great advantage to illuminate dark corners of the fossil record that did not permit palaeontologists to come to any satisfactory detailed conclusions. Sometimes the molecules simply helped to resolve difficulties of descent, revealing which species were most closely related at the molecular level and hence had most recently split from a common ancestor. Man's own descent was one of the first to be examined in this way, when Morris Goodman showed that as far as immunodiffusion was concerned there was nothing to choose between man and the African apes. This conclusion has since been strengthened at every level of analysis; protein fingerprints, amino acid sequences, DNA hybrids, nucleotide sequences (not for the whole human genetic blueprint, but for sizeable chunks of it), and every other technique available confirm that it is impossible to split the African threesome. On some measures gorilla and chimp are closer to one another than to man. On others man and chimp are closer, while on yet others it is man and gorilla who are nearest to one another. And for many of the techniques it is impossible to subdivide the three. As the data continue to accumulate it becomes inescapable that the ancestor of man was also the ancestor of chimp and gorilla, with almost no time during which any two of the three were together but separate from the other.

This information somewhat staggered the anatomists and palaeontologists, who had assumed all along, and enshrined their assumptions in taxonomy, that the great apes were all closer to one another than to man, but it wasn't entirely disastrous. To accommodate Goodman's findings they needed only to rearrange the primate family tree so that the line leading to man branched from the stock at the same time as the line leading to the apes. That way the two groups, apes and men, would have been separate for the same time and so would be biochemically close. Of course this didn't make too much provision for the poor orang-utan, so similar anatomically to the other apes and yet demonstrably further away from man at the molecular level. As far as they went, molecular methods seemed only able to confirm or deny the family tree that palaeontologists had devised from anatomical decisions. For some species the molecules produced the same tree as the palaeontologists, which made both sides happy, while for others they came up with a slightly different tree which meant that biochemists and palaeontologists alike had to pause for a moment and rearrange the branches of the trees. Molecules, it seemed, were a powerful tool for distinguishing between different lines of traditional evolutionary thought, and might even have a little new information of their own to contribute, but they would never amount to much more than that.

The realisation that set the new approach to evolution moving was that the molecular techniques need not be confined to a back-up role. The molecules contain *primary* information that can add dramatic new insights to evolutionary histories. That information is the molecular clock, which ticks off the years that elapse after two species have separated from their common ancestor. Learn to read the clock, and you have a view of evolution independent from, and additional to, the traditional disciplines of palaeontology and anatomy. 'I think there's an element of intellectual breakthrough in it,' says Allan Wilson. With micro-complement fixation as a clock, 'you've got the potential for a temporal framework for the whole of the living world. You don't have to rely on fossils any more . . . You can be testing theories about the mechanism of evolution in a quantitative way. I think that's the breakthrough that happened, and most people haven't realised that it even happened.'[8]

Credit for the concept of a molecular clock is essentially given to Emile Zuckerkandl and double Nobel Prize winner Linus Pauling. They are supposed to have been the first to realise that because mutations are random events DNA should accumulate them at a relatively steady rate. The process is very similar to the manner in which radioactive decay can be used as a measure of passing time, although there we usually talk about the radioactive 'calendar' rather than a 'clock'. In a way, the evolutionary molecular clock is a mirror-image of the radiocarbon, or potassium-argon, radioactive calendars. With radioactivity, half of the active atoms in a sample decay into stable atoms in some fixed time, called the half-life. The half-life of radioactive potassium is about 1,300 million years, and it is just under 6,000 years for radiocarbon. Successive halvings provide a measure of the time elapsed — the number of half-lives — since the radioactive material first formed. With the molecular clock, by contrast, the differences build up in the molecules. Two species descended from a common ancestor start out with identical DNA, and as the generations go by random changes accumulate in each species' DNA. The longer the species have been separated, the more different the DNA. The number of accumulated differences tells the time since the two species became evolutionarily distinct because mutation is a random process, as is the decay of a radioactive element, and while you can never say exactly which nucleotides will change, or when, you can say with some confidence how many of the millions of letters in the human genetic code will change in a thousand years, a hundred thousand years, or a million years. And that takes us the crucial step from acknowledging that the chimp and gorilla are our closest relations to setting a date on the split between our three species.

Zuckerkandl and Pauling are credited with the idea of the molecular clock, but their thoughts were put into practice by two young biochemists at the University of California at Berkeley. Vincent Sarich was then a graduate student doing a doctorate in anthropology. Allan Wilson was a biochemist who had just arrived at Berkeley, set up his lab and was about to get to work. Sarich's interest in the puzzles of primate evolution and anatomy has been fired by Sherwood Washburn, and after reading Goodman's work he realised that there was far more that could be done with the molecules. He went to see Wilson, and the two realised that the molecular clock was within their grasp. Sarich focussed his attention on two enormous blood proteins, albumin and transferrin, and measured the differences in these proteins in a large number of primate species. (And in the tradition of the great pioneer scientists, Sarich provided the first sample of *Homo sapiens* albumin himself.) All the primates are descended from a single common ancestor, and

looking at the data he could see that each primate lineage had accumulated the same amount of change since they split — a date reliably set
by the fossil record as 35 million years ago. The gibbons, African
apes, orang-utans, Old World monkeys, New World monkeys and
man each had the same number of differences in their albumin. So
ever since the lineages had split the molecular clock had kept on
ticking and the descendants of each lineage had collected the same
number of changes in the same length of time. This proved that the
molecular clock ticks — for a chosen molecule — at a steady rate in
related species. Other molecules besides albumin show the same
effect, although each molecular clock ticks at its own rate.
Cytochrome c, for example, changes by about one per cent of the
protein chain every 20 million years. Haemoglobin, by contrast,
shows a one per cent change after only 6 million years, an alteration
of one amino acid in each chain every 3½ million years or so.
Different macromolecules change at different rates, that much is
clear, but the key point is that the same molecule changes at the same
rate in all species studied. That simple observation makes the
molecular clock a reality.

Actually setting the clock is a simple matter. Choose a branching
in the family tree that is well dated by conventional techniques and
measure everything from this baseline. Sarich and Wilson decided
that one palaeontological fact of which they could be reasonably sure
was that the Old World monkeys split from the apes 30 million years
ago. (Sarich says now that this was a conservative estimate, and feels
the true date was probably closer to 20 million years ago; this does
not affect the argument significantly.) So they could count the tricks of
the clock in 30 million years quite easily, by comparing ape proteins
with monkey proteins; with that as a standard, they could set the clock
running and time the differences between man and the African apes.

The astounding result — published in 1967 — was that man, chimp
and gorilla shared a common ancestor no more than 5 million years
ago. In a perfect world a finding such as this, despite being so much
in opposition to the conventional wisdom, would have led to the
palaeontologists examining their own case to see if it wasn't possible
that Sarich and Wilson were indeed correct. Nothing of the sort
happened, and the response was by and large not very favourable.
The traditionalists raised objections that showed a fundamental lack
of understanding of the molecular methods and a wilful distortion
of Sarich's position. To pick a comparably significant development

at around the same time but in another branch of science, it is as if theoretical astronomers had ignored the discovery of pulsars. We will return to some of those objections shortly, but for now it is important to note that the very first species on which the molecular clock was set to work was our own, and the very first timings that it produced were extremely disturbing. No wonder this new–fangled technology, with its patently absurd answers, was given such short shrift. How very different things might have been if Sarich had been less interested in man and more concerned with other, less important, species. He did go on to extend the methods to other species, as we will now see, and answered many a problem that had puzzled conventional palaeontologists, but it seems that in the eyes of the palaeontologists he can never make up for challenging the human fossils at once, going first for the biggest prize of all.

After Sarich's new timescale for primate evolution met with a sceptical response he turned his attention to other species to establish the validity of the molecular clock technique. The first group he studied was the pinnipeds. These are the seals, the sea–lions, and the walrus, mammals that have returned to the water but breed on land. All pinnipeds are carnivores, but the exact details of their evolution are not easy to glean from the fossil record. Nevertheless, based on the fossils and on contemporary anatomy it is usual to group the sea–lions and the walrus together into one unit (the most obvious shared characteristic is that sea–lions and walruses have ears, whereas true seals do not) which split off from the ancestral line of canids (dogs) in the middle of the Miocene, about 20 million years ago. True seals, on this picture, form another group which is more closely related to the so–called mustelids (weasels, otters, badgers and so on), from which they split at about the same time, 20 million years ago. The mustelids split from the canids about 35 million years ago in the Oligocene, before the accepted date that the pinniped ancestors had begun their return to the water, so that by this account the remarkable similarities of the seals and the sea–lions are nothing more than coincidence, the inevitable result of a carnivore returning to the water. As noted University of Reading palaeontologist Beverly Halstead puts it, 'the adaptations both groups have made are the result of convergent evolution'.[9] Other authorities agree that the eared seals are descendants of a different stock to the true seals, but feel that the sea–lions came from a bear-like ancestor while the true seals came from an otter-like ancestor. Either way, the similarities are seen as coincidences.

Sarich decided that this was as good a test of the molecular clock as any. The first step was to obtain the albumins and make the antisera. In this, like Nuttall before him, Sarich is dependent on the good will of friends and colleagues. He collected cat and mongoose sera from the University of California, raccoon from Wisconsin and black bear from the University of Puget Sound. Harbour seal and California sea-lion were provided both by the Stanford Research Institute and the University of Washington, while the San Diego Zoo provided a specimen of night monkey blood and the San Francisco Zoo donated blood from the binturong, a strange animal that looks like a raccoon but is in fact a member of the cat family. Finally, with the purchase of some mink locally, Sarich had representatives of all the animal groups he needed. He prepared antisera to the primate (night monkey), pinnipeds (harbour seal and California sea-lion), canoids (dog, bear, raccoon and mink), and feloids (cat). He tested all these against one another and also against the walrus (another pinniped) and the binturong and mongoose (feloids). The results were absolutely straightforward. The immunological distance between the pinnipeds and canoids was about 43, that between the pinnipeds and feloids about 87, and that between the canoids and feloids also 87. There was, as he reported to the scientific community, 'a two-fold greater similarity between the albumins of the pinnipeds and canoids than between either of these and those of the feloids'.[10]

On its own, however, even this discovery could not untangle the complicated background of the pinnipeds. On the face of it the conclusion was simple; pinnipeds were descended from the same line as dogs, about half as long ago as dogs and cats diverged from their common ancestor. But this would be true only if albumin had been evolving steadily in the carnivores, as it had in the primates. If this was not the case, then Sarich could easily have said that the seals were evolved from cat-like ancestors, and that after this happened the albumins of the cat group suddenly began to change at twice the normal rate. Fortunately, when he compared the carnivore albumins with primates (represented by the night monkey) Sarich discovered that each had accumulated the same number of changes; they had indeed been evolving steadily, and could be used as a clock. This being the case, there is only one evolutionary history that fits the facts. 'Pinnipeds are derived from a canoid stock,' Sarich asserted, and 'shared a long period of common ancestry with the canoids after their divergence from the feloid line'.[11]

So far, so good. The pinniped tangle was beginning to yield. But what about the further questions of just when the line split from the dogs, and how the various pinnipeds are related? These, too, are answered by the molecules. Simply drawing the evolutionary tree with the branches as long as the immunological distances provides a great deal of information. Cat and dog, by this reckoning, split an arbitrary 49 time units ago. On the same scale, dog and seal did not split until 24 units ago. 'This conclusion is compatible with fossil evidence,'[12] Sarich concedes, in that there are fossil feloids and canoids from the Eocene, some 40 million years ago, whereas the first pinniped is not known until about 18 million years ago in the Miocene. But even so, and in the absence of any positive evidence, many authorities, including the leading mammal palaeontologist of the time, Alfred Sherwood Romer, felt that the split between canoids and pinnipeds occurred just after the split between canoids and feloids. As to the internal relationships of the pinnipeds, Sarich elucidated these too. He demonstrated unequivocally that, contrary to accepted dogma, all pinnipeds were derived from a single common ancestor, one that they shared with the ancestor of the dogs, and went on to time the splits between the various pinnipeds.

Satisfied that the evidence for clock-like behaviour in carnivore albumins was every bit as good as in primate albumins, Sarich first set the carnivore clock. The primates had shown that an immunological distance of 100 units is equivalent to a separation 60 million years ago, so the separation of the cat from the canoid carnivores could be put at 87×0.6, or 52 million years ago. The same conversion factor puts the split between the pinnipeds and the canoids at about 28 million years ago and the divergence between the two pinniped groups in the middle of the Miocene, about 15 million years ago. There are very good fossil pinnipeds from this time, fully adapted to life in the water. This effectively rules out any notion that the eared-seals are derived from a different terrestrial ancestor than the true seals. They split from a common seal stock, only 15 million years ago.

Within the pinnipeds, it seems that the walrus and the California sea-lion have been separated for only 5 million years, much later than palaeontologists had previously thought, and coincidentally at the same time on the molecular clock that man's line split from the

African apes. If walruses — or sea-lions — could write books and contemplate their ancestry, how might they have received these revelations?

Another animal to come under Sarich's molecular microscope is arguably the most well-known animal of all, especially since it was adopted as a symbol 20 years ago by the World Wildlife Fund. Rare and mysterious, living only in one small part of the globe, and with a face that could have been designed specifically by a manufacturer of cuddly toys, the giant panda is something of a contradiction. Everyone can recognise it, but we actually know very little about it. Aside from all the other mysteries of its life in the wild there is the fundamental mystery of its zoological classification; it seems closest to either the bears or raccoons, but controversy surrounds its exact status. Three positions are common. Some say the giant panda is closest to the lesser panda; this is clearly a raccoon so the giant panda should also be considered a raccoon. Others say that the two pandas are indeed closely related, but that the lesser panda is not a raccoon; the two pandas should be in a separate family of their own. Finally, there is the view that the giant panda is nothing but an aberrant bear, and so should be in the Ursid family along with the other seven bear species; this says nothing about the status of the lesser panda. Now whether the giant panda is a bear, a raccoon or a panda may not unduly worry the man in the street or the World Wildlife Fund, but it is clearly an embarrassment to zoologists. And while it obviously worried the man in the street and the World Wildlife Fund a great deal when London Zoo's own giant panda, the ever popular Chi-chi, finally passed away in June 1972, one zoologist, Vincent Sarich, was secretly rather pleased. In no time at all a sample of Chi-chi's blood was on a plane from Heathrow to San Francisco and Sarich's Berkeley lab, where it would be used to seek a resolution to the panda's identity crisis.

A little more than a year later, the journal *Nature* carried the answer: 'The Giant Panda is a Bear' proclaimed the headline over Sarich's paper.[13] He had extracted the albumin and transferrin from Chi-chi's blood and from a number of other animals of interest, including the lesser panda, some bears, and several raccoon family members. Using the by now familiar immunological techniques he measured the distance between all these species and the antisera of the black bear and the raccoon. The first interesting point to emerge

(109)

was that the bears and raccoons are rather close to one another, certainly closer than any other two families among the carnivores; the distance between them was 97 units, while for other pairs of carnivores it ranged between 110 and 150 units. The bears and raccoons thus shared a period of common ancestry after they had split from the other carnivores, which means that the whole question of the panda's position becomes a little easier to understand; if the bears and raccoons are so similar, no wonder it was difficult to decide whether the giant panda was a bear or a raccoon. More to the point, though, the distance between giant panda and black bear turned out to be some 22 units, closer by far than the distance to the raccoons. The panda was a bear, and quite recently too, for this distance implies a splitting time of about 7 million years ago, a little before the main radiation of the bears but still quite recent compared to the 25 million years of the raccoon–bear split. The immunological data also showed up a discrepancy concerning the lesser panda. This appealing animal, which does indeed look rather like a raccoon, is no more a raccoon than the giant panda. It is separate from the raccoons and from the giant panda and other bears, and forms a group of its own, probably descended from the bears but about 18 million years ago as an early offshoot of the bear line.

Sarich's conclusion, that the giant panda is a bear and that the lesser panda is neither a raccoon nor a bear, but is slightly closer to the bears, agrees with the one comprehensive anatomical study that has been made, but conflicts with the story of carnivore evolution as it is usually told. Once again, there is no fossil evidence that contradicts the molecular view, but still the molecular evidence has not found ready acceptance. Nor are carnivores the only group that has been investigated in this way. Molecular evolutionists have now looked at almost every group in existence, using their techniques to sort out confusion and solve mysteries for which fossils provide no clues, and their answers should by now be an accepted part of the database on which evolutionary trees grow. Sarich noted somewhat ruefully in an early paper on pinnipeds that 'many workers have been hesitant in accepting phyletic conclusions based on comparative studies of protein structure; such hesitancy should no longer be necessary'.[14] Indeed it should not. To be frank, though, the molecular data have not been accepted as readily as they might, even when the species involved is not our own. The reasons for this are many and varied, and usually more to do with gut-feelings than science. In our search for the story behind this revolutionary development in science, we have lost track of the number of people who have replied to our questions with a vague and uneasy 'Well, I don't know what's wrong with it, but I feel that molecular evolution isn't quite right.' There is nothing wrong with molecular evolution, but ever since Sarich and Wilson claimed that man had separated from the apes less than 5 million years ago their results, and those of other workers in the field, have been mistrusted. No matter that the date they advanced is not contradicted by a single fossil, or that the palaeontologists are slowly and surreptitiously making man younger themselves: the explosion of that first bombshell is still reverberating through the halls of palaeontology. The time has come to examine the explosion, and the damage it did, in detail.

———— Five ————

THE RULES OF THE GAME

'We suggest that apes and man have a more recent common ancestry than is usually supposed.'[1] That simple and innocuous sentence, two-thirds of the way through the summary of a paper, sums up the explosive force of molecular anthropology. The paper was by Vincent Sarich and Allan Wilson, and it appeared in *Science,* perhaps the most widely read scientific journal in the world. The date was 14 September 1967, and quite obviously the world was not ready for this kind of revelation, for the palaeontological orthodoxy simply didn't believe it. Their objections, as we shall see, in fact carry very little weight, but we must confess that at the time they were probably right to challenge Sarich and Wilson. After all, here was a new technique, untried before, being used to address the *big* questions: Where did man come from? And when? Admittedly Sarich had a joint appointment in the Berkeley departments of anthropology and biochemistry, but in essence here were two chemists telling the glamorous fossil-hunters their business. It simply could not be correct. Or could it?

So important is this one result, and so misunderstood are the methods behind it, that we think it appropriate to go carefully through the calculations to show exactly how Sarich and Wilson arrived at their iconoclastic conclusion. We have already seen how the immunological methods, honed and refined into a delicate tool, provide a measure of the distance between any two species. The primary observation in the *Science* paper was that the hominoid (apes and man) albumins were much more similar to each other than any of them was to any nonhominoid albumin.

There are two possible explanations for this. On the one hand, it could be that since the apes and man split, the rate at which albumin accumulated changes has slowed down considerably in either, or both, lines. This would mean that, as far as their albumins were concerned, apes and men were close because their albumins hadn't changed much. On the other hand, as Sarich points out, 'the close molecular relationship may reflect a more recent common ancestry between ourselves and the living apes than is generally supposed, albumin evolution having proceeded at the usual rate for primates'.[2] As we said in Chapter Four, Sarich had already looked at the albumin changes of all the primate lineages, and had discovered that each line had collected roughly the same number of changes. That being the case, albumin must have been evolving at the same rate in apes and men as in monkeys, which tends to make the first explanation rather unlikely. Apes and men are similar not because they are evolving more slowly than monkeys, but because they have had little time in which to evolve away from each other. That much the albumin molecules have already told us; they will also reveal just how little time there has been.

Because all the primate lines have accumulated the same amount of change to their albumins, there must be a calculable relationship between the time since any two species split and the immunological distance between their albumins. 'This relationship is likely to be rather simple,' say Sarich and Wilson,[3] and indeed it is. Investigation of the immunological distances for a set of proteins from fishes, amphibians, reptiles and birds gives a simple equation that relates the time of divergence to the immunological distance divided by some constant. The constant can vary from group to group and takes into account the fact that some proteins evolve at different rates in different groups; to discover what the constant is you need to have good fossil evidence for the date of one particular split, and then everything can be related back to that date. The one split that Sarich and Wilson felt they could date with some certainty was that between the Old World monkeys and apes and man. This, they said, probably took place 30 million years ago, and although the fossil evidence for this was a lot scantier than the evidence for fishes, amphibians, reptiles and birds, there was certainly nothing to contradict it at the time. The immunological distance between six species of monkey and the apes was 2.3, which after plugging into the equation gave a value for the constant of 0.012. Now that they had the value of the constant, Sarich and Wilson could use this in their equation and plug in the immunological distance between man and the African apes; the result was that the three species emerged from their single common ancestor 5 million years ago. Exactly similar calculations for the orang-utan (ID 1.22) and the gibbon (ID 1.28) revealed that these split from the man-African ape line 8 and 10 million years ago respectively.

And there it is. Simple calculations reveal that man and the apes share a very much more recent ancestry than is usually supposed. There might be errors in the calculations, perhaps because the benchmark date, the split between the Old World monkeys and the hominoids, was in error, or because the equation relating time of divergence to immunological distance was not quite right, but given the very clear grouping of man, chimpanzee and gorilla into one immunological unit it seemed a remote chance that any changes in these factors would lead to any radical revision of the date given. 'These possible errors are unlikely to be of sufficient magnitude to invalidate the conclusion that apes and man diverged much more recently than did the apes and Old World monkeys,' was how Sarich and Wilson put it.[4] And of course this very recent split solved at a stroke many of the problems that we have referred to as the monkey puzzle. If man and the apes had only recently separated, then their remarkable anatomical similarities were not the result of coincidence but instead were the heritage of a common ancestor.

Sarich and Wilson did not simply stop work after their 1967 paper. They continued with their exploration of molecules as witnesses to evolution and, as we have seen, were able to solve palaeontological problems that beset other groups of species. Nor was their own work ignored by all. Biochemists and molecular biologists took up the challenge and looked for more evidence of the similarity between apes and men and, in particular, for other clocks that might be used to supplement, or discredit, the albumin clock. Immune responses are, as we have already noted, some way away from the sequence of bases along the DNA; that does not alter their usefulness, but it does provide something of a sticking point when trying to persuade reluctant palaeontologists, who might (though we doubt it) be more agreeable if DNA itself provided the same outcome. While nobody has sequenced the entire genetic blueprint of man, chimp, or any other mammal, there is more than enough information available to use a DNA clock. In point of fact it is not the actual sequence of nucleotides that we will use here, though that, too, can be done, but DNA hybridisation data, which in a sense is better than the sequence of one particular gene because it looks at the entire genetic blueprint and compares species using all their DNA.

Raoul Benveniste and George Todaro at the National Institutes of Health outside Washington D.C. have created DNA hybrids between four representative primates.[5] They looked at man, as a hominid; gibbon, as a hominoid; baboon, as an Old World monkey;

and the squirrel monkey, as a New World monkey. When the DNA melting points had been translated into percentage differences, they found that man and gibbon differed by 6 per cent, man and baboon by 9 per cent, and man and squirrel monkey by 15 per cent. The first step is to convert these figures into some sort of evolutionary tree. Starting with the 9 per cent difference between man and baboon, we want to know how much of that difference was the result of evolution in man's line and how much was in the baboon's line. We know from comparative anatomy and other molecular work (there are no relevant fossils) that man and the baboon shared a common ancestor for some time after the squirrel monkey's line had split off, so if man and the baboon have evolved at different rates from one another since they split, this will show up in the form of different distances between each of them and the squirrel monkey. Any difference between the man to squirrel monkey distance and the baboon to squirrel monkey distance can only be the result of a disproportionate change in either man or baboon. But in fact, when one looks at the melting points of the hybrid DNAs, one finds that the man to squirrel monkey distance and the baboon to squirrel monkey distance are both equal, both 15 per cent. So there has been no differential evolution between man and the baboon and we can quite justifiably apportion the 9 per cent equally between them, 4½ per cent in the baboon line and 4½ per cent in our own. The amounts of change are not *assumed* to be equal, they are measured and *proved* to be equal.

Now we are ready to introduce the gibbon. The distance between gibbon and man is 6 per cent, the gibbon to baboon distance is 9 per cent, and the distance to the squirrel monkey is again 15 per cent. This last figure means that gibbon, like man and baboon, has changed at the same rate relative to the squirrel monkey, so we can again split the difference between man and gibbon into 3 per cent along each line. But we have already worked out that man accumulated 4½ per cent differences since he split from the baboon; so the gibbon must have diverged from man after a common period in which the two lines accumulated changes in 1½ per cent of their DNA.

The final picture is unambiguous, and what takes many words to explain is easily seen in the diagram. We have, using only molecular data, created an evolutionary tree for four major primate groups, and it doesn't matter whether the molecular information consists of immune differences, nucleotide sequences or anything in between;

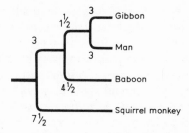

A simple molecular tree. The percentage DNA differences have been apportioned to the various branches on the road to each species as explained in the text.

provided that there has indeed been regular evolution, and this is easily checked by looking at the numbers, we can apportion the evolution between the various species and construct an evolutionary tree. There remains only the business of setting a clock to time the growth of the tree. With immune distances the clock can be provided by the now enormous collection of immune distance data points in combination with one or two well-dated splits that can be obtained from the fossil record. The same is true with the other methods, but it is important to note that the tree that we construct, with its differences shared out appropriately between the lineages, does not itself provide absolute times; it only provides an estimate of the relative time since the splitting of several lines.

Immune distances, for example, show that, compared to the distance between man and chimp, man and baboon are 4½ times further apart, man and squirrel monkey 8 times further apart, and man and lemur 16 times further apart. In other words, the times since these five species separated are in the ratio 1:4:5:8:16; on its own, the immunological information can never tell us whether the actual times are, say, 2, 9, 16 and 32 million years, or 20, 90, 160 and 320 million years. 'The molecules are totally impartial in this regard,' says Sarich.[6] The calibration must come from the fossil record, supplemented where necessary by common sense. For example, the second set of figures is extremely unlikely, because 320 million years ago, when the lemurs would on this picture be splitting from the rest of the primates, the amphibians had only just crawled out of the Devonian swamps and there were no mammals, let alone primates. The fossil record may be incomplete, but we do know enough to ensure that this kind of absurdity does not slip through, and when all the evidence is taken into account the gap into which we have to fit the molecular dates is rather narrow, and the dates correspondingly accurate.

Molecules of all sorts have now been put through their paces to illuminate the evolutionary history of the primates — indeed it is yet another manifestation of our self-centredness that the primates are far and away the best studied group — and all of them agree, not only on the closeness of what we call the African triumvirate but also on the recent emergence of that triumvirate. On almost any measure you can think of, man, chimp and gorilla are closer to one another than to any other living animal. Their DNA, where that has been looked at in detail, is all but identical. Their smaller proteins, which have been completely sequenced, are often identical, and where they do differ it is generally by less than one amino acid in a hundred. Larger proteins show great similarities too, and in 1975 Allan Wilson used all the information he had to calculate the average difference between the species. Thirteen proteins, ranging in size from 30 to 650 amino acids long, differed by a total of about 19 amino acids. Averaged over the entire 2633 amino acids that works out to 7.2 differences per 1000 amino acids: the proteins of man and chimp are more than 99 per cent the same.[7] Our chromosomes are practically identical too. Recall that all chromosomes come in pairs; man has 23 pairs, 46 in all, and the chimp has 24 pairs, and the difference seems to be that the pair labelled No. 2 in humans is a joined version of two pairs of chromosomes that are separate in the chimp, but the detailed pattern of light and dark bands that can be seen on a suitably stained chromosome is practically identical. As far as can be made out, there are just nine regions that differ, three of which only became visible with the advent of new staining techniques in 1979, and in these regions it looks as if a portion of the chromosome has been turned over in one of the two species.[8] This does happen occasionally (Chapter Three) and may have a profound effect on the development of the animal concerned, but there is no way of telling whether the chimp's chromosome is the right way round and ours inverted, or ours is the original and the chimp's an inversion. Perhaps four or five inversions from some common set produced man, and a different few produced the chimpanzee.

Where the data permit a clock to be used, the conclusion of the 15 years of work since Sarich and Wilson's first discovery is that the three of us split from one another about 4½ million years ago. This

(117)

may be a bitter pill to swallow, not only for the palaeontologists upon whose territory it poaches, but also for many uncommitted individuals who would nevertheless like to put as much time as possible between themselves and those laughable animals the chimpanzees. And, as is to be expected, people find many good reasons not to swallow bitter pills. We would not be fair to the palaeontologists if we didn't voice some of their objections here, and indeed it suits our purpose rather well to do so, for we hope to show that they have no good grounds for those objections. Before we get to that, though, there is one final piece of evidence that has very recently come to light. It is not, by any means, the last word, just another in the ever growing mountain of results that all support one another and simply cannot be ignored.

Allan Wilson and his colleagues have been looking in detail at the genetic make-up of the tiny organelles called mitochondria.[9] These are present in every living cell and their function is to provide the cell with energy, which they do by taking the energy from the bonds holding food molecules together and transferring that energy to so-called high energy phosphate groups, which can then be shunted around the cell to wherever they are needed. Mitochondria have their own DNA, and reproduce themselves more or less autonomously within the cell, just two of the surprising facts that have led many people to consider them either as a slave or a parasite, having long ago forsaken life as an independent bacterium-like object for a cosy niche inside other cells. There are three interesting things about mitochondria for our purposes. Firstly, they have their very own DNA, with its own, slightly different, genetic code; secondly, mitochondria are passed on to the next generation solely by the mother, who hands on some of her mitochondria in the cytoplasm of the egg; and thirdly, the rate of evolution of mitochondrial DNA seems to be ten times faster than that of normal DNA in the nucleus. These three things mean that a study of the molecular evolution of mitochondrial DNA is not tainted by even the slightest suggestion of generation times or advanced development, objections, as we will see, that have been unsuccessfully levelled at other studies. This is because the mitochondria divide every time a cell divides, regardless of whether sexual reproduction is imminent or not. And the rapid rate of change means that the very close spacing of man and the apes can be spread out and examined on a larger scale map, with a corresponding increase in the accuracy with which we can measure the distances between them.

The answer from the very recent studies of mitochondrial DNA will not come as a surprise; man, the chimp and the gorilla diverged, according to these molecules, sometime about 4 million years ago. And even on the large-scale map it is difficult to separate man, chimp and gorilla. There seems to be a small amount of evidence that chimp and gorilla shared a common ancestor for a very brief period after they split from man, but it is by no means conclusive, and other evidence favours an association between man and chimp after they split from gorilla. Mitochondria will also have some very exciting things to tell us about the later stages of human development, long after the split between man and apes, as we will see, but having presented just a small portion of the evidence in favour of the molecular clock we should now look at the objections.

One very simple objection is that because the clock measures changes in the DNA, and because these mutations occur largely when animals form sex cells, the rate of evolution would be dependent on the speed with which animals reproduce. A mouse for example, which produces a new generation every three months, would accumulate far more mutations than the much longer lived, and slower breeding, bat. Or, to come closer to home, a little rhesus monkey, which reaches maturity in three or four years and so has a generation length five times shorter than our own, might evolve, or accumulate mutations, five times faster than man. This is, on the face of it, a reasonable objection, but it isn't enough just to mention it and then do nothing. It makes a very clear prediction, and the way of science is to test predictions and, if they are not fulfilled, to throw out the hypothesis that made them. What happens when we look at the rate of evolution, as measured by immunological techniques, for example, and compare it with generation time? There is no effect. Sarich measured the immunological distance between all the primate groups and a representative carnivore; the two furthest away from the carnivore were man, 162 units distant, and the rhesus monkey, 166 units distant. The rapid reproductive rate of the rhesus monkey has not had any effect on the rate at which it has accumulated mutations, and in fact there are many primate species that are closer to the carnivores and have even shorter generation times than the rhesus monkey. The hypothesis that generation time affects rate of evolution must be discarded, and won't do as an objection to the molecular clock.

So it seems that instead of taking place only when a cell divides to form a gamete, or sex cell, which would tie the rate of mutation to the generation time, mutations tend to occur regularly through an organism's life. They are related to real time, not to reproductive time. This very phenomenon, that mutations seem, according to the evidence of the molecular clocks, to accumulate at a steady rate, is deeply disturbing to many biologists. The reason for their concern is that it implies that the mutations are selectively neutral, that is that they are neither good nor bad for the organism carrying them. While this may be true in many cases, it should not be seen as a block to acceptance of the molecular clocks, but rather as a spur to greater research efforts. We will never understand the regularity of mutations if we deny their existence. Nobody has even begun to look at the problem of why some mutations are regular events, or why they are not restricted to the simple errors in copying that accompany cell division. It might be that the cell's repair mechanisms are even better in sex cells than in the ordinary cells, so that most of the errors that are transmitted are the result of, say, cosmic rays rather than bad copying. But, it does not really matter what the mechanism – or indeed mechanisms – may be, for we do not have to understand a phenomenon to make it work for us. As Sarich is fond of pointing out, nobody yet understands the ultimate basis of gravity, but that doesn't stop us sending men to the Moon or space-probes to Saturn.

Another frequently voiced objection is that if the rate of evolution, the ticking of the clock, were the same in all species, then the rate of visible evolution would be the same. Man and chimp look so different, while chimp and gorilla do not. This is an important point, one that we will return to later when we discuss in more detail the changes that enabled the rapid evolution of man, but as an objection to the clock this too will not hold water. The placental mammals have been around for about 100 million years, and only really spread out in the past 65 million years, since the great dinosaur extinction; not surprisingly, perhaps, there are huge differences between, say, a blue whale and a bat. But frogs have been around considerably longer, perhaps as much as three times longer, and yet one species of frog is very much like another. In fact, so diverse are mammals that taxonomists put them into at least 16 separate orders, yet they place all 3,000 or so species of frog into just one order. Nor is this simply a case of mammals looking different to us because we are mammals, while frogs all look the same because we are not frogs. Unlike the man who thinks all Chinese are alike because he hasn't yet learned to see the differences between them, frogs really are all alike; certainly they are much more similar than mammals. A huge series of measurements made by Allan Wilson and two colleagues, Lorraine Cherry and Susan Case, demonstrated clearly that chimps and men, investigated using the very indices that taxonomists use to classify frogs, are indeed more different in shape than even the most distantly related frogs. And yet, of course, genetically they are closer by far than the closest frogs.[10]

So here we have a case of a short period of evolution, exemplified by the mammals, producing enormous change, and a much longer period, for the frogs, producing much smaller change. The way out of this impasse harks back to the difference between structural genes and regulator genes, and the fact that a mutation in a regulator gene can have very profound effects on the appearance of a species. It takes only a very small change in a regulator to affect a whole range of structures, and although DNA relates directly to the structure of proteins there is no such link between the proteins themselves and the gross structure of the animal. In point of fact when you look at the DNA for bat and whale you find that the differences are no larger than they are for two apparently closely related species of frog. Appearances are deceptive, at least when it comes to molecules and morphology, and we predict that if biochemists ever learn to isolate regulator genes then that is where they will find the differences that signify blue whale or bat. The fact that man and chimp look so different, and chimp and gorilla so similar, is a red herring, not even a problem for molecular evolution or the regular ticking of the various molecular clocks.

It is possible, of course, that in the examples we have just looked at the fast evolution that is supposed to go with quicker reproduction exactly counteracts the slower evolution that is supposed to go with less outward change. This, again, is easily tested. As Sarich explains, 'we simply set up a comparison where the variable is either degree of organismal change or generation length, but not both'.[11] The comparisons he chose were man and rat, which are both highly evolved forms that show a great deal of outward change from their respective ancestors, and yet have very different generation times, and tree shrew and rat, which both have short generation times but differ in how much they have evolved from their forebears. The three species were then measured against two outside reference points — a carnivore and a bat — for good measure. It turned out that each of the three — man, rat and tree shrew — is the same distance, about 160 units, from the reference species. 'Thus,' Sarich says, 'neither generation length nor degree of organismal evolution are having an effect on our *observed* rates of albumin change. These are just illustrative but representative examples of the general picture — simply stated, the correlation between either degree of organismal change or the generation length and the amount of genetic change is zero.'[12]

Morris Goodman himself, who started this evolutionary ball rolling in the modern era, prefers to accept fossil inference rather than molecular evidence. He believes that the clock runs at different speeds at different times and in different lineages, and that while it is all right for decisions of who split from whom it is no good at all for deciding when that split took place. By his own reckoning there have been periods in evolution — for example 400 million years ago when vertebrates began their conquest of the land — when molecular changes to the DNA were 10 times more common than they are today. Goodman claims that what looks like a clock is no more than an averaging, over hundreds of millions of years, of stop-start changes that make no sense whatsoever at a smaller timescale. As a protein becomes adapted to a new function, for example as haemoglobin became adapted to the job of picking up and offloading oxygen, so it becomes more vulnerable to change and more resistant to mutation. The rate of molecular substitutions, according to Goodman, slows down as a lineage becomes more advanced.

It is true that some molecules do not behave as clocks, and do not accumulate changes on anything like a regular basis. Goodman's ideas do apply, for example, to the major changes that have shaped haemoglobin, and if his objection were universally true it would indeed be cause for concern. Once again, however, it is perfectly possible to measure the rate at which a molecule accumulates changes, and in fact this was done in the very year that Goodman first voiced his objections. W. O. Weigle showed clearly that human albumin was just as different from that of the cow as were the albumins of rat, mouse and dog. They had all changed equally from their common ancestor, despite some being more advanced, in Goodman's terms, and of very different generation times.[13] Of course if a molecule — haemoglobin, say — doesn't behave like a clock there is very little point in attempting to use it as a clock. 'We wouldn't be dumb enough to try it,' Sarich assured us.[14] He also wonders why it is that Goodman has focussed his attention on haemoglobin rather than on the albumins and transferrins that provide so much of the published clock timings. Furthermore, Sarich and Wilson have compared advanced and primitive lineages directly. They looked at eight different proteins in the various primate lineages, counting the number of changes accumulated in each lineage. They don't need to know when the lineages split, because all they are asking is whether the mutation rate in the more advanced lineage, man's, is slower than the rate in a less advanced lineage, the baboon's. The answer is a categorical No. As they said at the time, 'the results give no support to the idea of a molecular evolution slowdown specific to the lineages leading to humans and African apes,' which is what Goodman would have us believe.[15] Goodman says 'I do not reject paleontological evidence of ancient splitting times' — but there is no good fossil evidence for him to accept or reject when it comes to the date of the man-ape split.[16] So why not accept the only good evidence there is, the molecular clock? As Allan Wilson points out, Goodman shows 'a touching faith in what the paleontologists have to say'.[17]

The most common objection really encompasses the three we have just looked at. It says, simply, that the molecular clock's timings are no good because you have to assume steady rates of evolution. This is completely untrue, and anyone who seriously believes this has quite obviously not read any of the relevant papers, or at any rate has not understood them. Wherever molecular information is used to provide timings, rates of evolution are measured, not assumed; if they are not steady then there is absolutely no point in attempting to use that molecule as a clock. Sarich is quite bitter about this. 'It has been the refusal to recognize this point . . . and consequent rejection of our efforts by continual reference to our "assumption of equal rates of change" that has been the most disturbing and depressing aspect about the reception our work has received,' he says. 'It may be that we are so conditioned to think of rates as equal to amounts divided by time that we forget that rates over equal lengths of time can be compared without knowing the times involved, as they cancel out in the comparison. All we need to know is that the two lines have existed for the same length of time, and this is an evolutionary given — any pair of living species have existed for precisely the same length of time since they separated from their common ancestor. Perhaps this is simply too obvious to have been noticed; all I know is that the point has been almost impossible to get across — whether the audience be a group of undergraduates in an introductory human evolution course, or experienced workers in the comparative biochemistry of proteins and genes.'[18] The surprising thing, perhaps, is how many very good, clocklike molecules there are, and how often the rates of evolution are in fact steady. Occasionally one finds anomalies, as would be expected in any process like mutation that is not entirely predictable, but these can be accounted for and often used to very good effect. The objectors, then, do not have a case.

Perhaps the best evidence in favour of a clock is what happens when you assume that the clock is true and then see what that tells you about evolution. We treat the clock as a hypothesis and use it to predict what we will find when we measure the fossil evidence against it. A look at 7 proteins from 17 different species, from rat to man, gives practically a straight line of molecular change against fossil dates. That straight line means that there *is* a clock, a regular relationship linking the immune distance between two species on the one hand and the time since they split on the other. We can go further with such a graph, as Sarich and Wilson have done, and add to it the best fossil data that we have, with two separate pieces of information for each split. One is the minimum possible date for the split. This, obviously, is the earliest date at which we can find examples of both lineages in the fossil record, or if not both then one of the lineages showing some clearly defined feature that must be an evolution from the common ancestor. The other is the maximum date for the split, which is given by the date of origin of the group itself; a group cannot split before it exists. These two limits, upper and lower, can be fixed with reasonable certainty for a number of splits that can also be measured biochemically, and they define a series of windows through which the line of the molecular clock must pass. The split between the true mammals and the marsupials, for example, cannot be more recent than 100 million years ago, because after that date we have fossils of both types. Nor can it be more than 125 million years ago, because we do not have any good evidence of the mammals before that date. Just so with primates, which must have split from the mammals between 95 and 75 million years ago. By 60 million years ago we have evidence of the prosimians — lemurs, lorises and tree shrews — in the fossil beds, so this is the latest date for that split, and the same goes for the Old and New World monkeys; they must have split between 55 and 35 million years ago. Fossils of apes mean that this group had split from the Old World monkeys by 30 million years ago — but then we enter the most difficult patch. There are, as we've said repeatedly, no fossil ancestors of apes that are not also ancestors of humans, so in principle we might have split very recently indeed. But there are remains — Lucy and whoever made the footprints at Laetoli — that show a primate walking upright 3.6 million years ago. This is clearly a feature derived from a more primitive four-footed type, so the split between man and the African apes had almost certainly occurred by this time.

Even this is not certain, however. Perhaps our conception of the change from a four-footed gait to upright walking being a unique feature of *human* prehistory is a misconception left over from the outmoded idea of a ladder of evolutionary progress, with man at the top. It is entirely possible, in evolutionary terms, that about 4 million years ago an ancestral ape species 'learned' to walk upright and produced the famous fossil footprints in the lava plains that are now exposed at Laetoli in East Africa, but that the split which produced the three modern forms — man, chimp and gorilla — occurred *after* this significant development. Certainly chimps do not walk upright today, or when they do it is occasional and ungainly, but evolution is neither a ladder nor a one-way street, and modern chimps could be descended from ancestors who walked upright. Such a change would, after all, be far less drastic than the changes that turned an ancestor of the dogs into a plethora of sea-going mammals. We do not suggest that there is any proof that chimps share with man an upright ancestor and that the chimp line 'forgot' this vital trick (indeed, on balance the anatomical evidence is against it). But we do wish to point out that such a possibility conflicts neither with the fossil evidence nor with evolutionary theory. All too often even respected palaeontologists seem still to think in terms of a ladder of evolution with man at its top, implying that any improvement in the ancestral line must be fixed for once and for all. The *possibility* that the other African apes branched from our shared line only after it became recognisably human is worth entertaining as a mind-broadening exercise. If, however, we were asked to sketch a 'most likely' picture of human evolution, we would certainly be more inclined to comment on the remarkable 'coincidence' that the first evidence of upright walking in the human line comes just after the date of the man-apes split as revealed by the molecular clock. The two pieces of evidence, taken together, suggest that it may well have been his upright posture that first set man apart from the African apes some 4 million years ago (see Chapter Six).

We have, then, six well established windows on the fossil record of our evolution, and when we fit the windows provided by the fossils to the graph of molecular distance against time, we find that the line passes beautifully through each and every one. At no point does the molecular clock disagree with the fossil evidence. Parts of the palaeontological record may be poor, but those fossils that we do have provide very good splitting times for certain lineages. They form a series of windows that the clock must pass through; it does.

Each separate bit of molecular evidence, for one group of species or another, only provides, as we've said, a ratio of splitting times. The actual times are set by reference to accurately dated splits from the fossil record, but changing any one of those timings is bound to have repercussions on all the other timings, because the evidence is that rates of evolution are steady even in different groups. You just cannot have one set of timings for one group and another set for another group, because there is solid data showing that the molecular timings are all interrelated. So, for example, rather than have a man–chimp split about 4½ million years ago, based on a well documented split between man and the New World monkeys some 35 million years ago, the palaeontologist is perfectly free to put the origin of the human line back where it was once assumed to be, at around 20 million years. But then he must accept that the New World monkeys split off 160 million years ago, long before the mammals appeared on the face of the Earth. Inescapably, the molecular clock does exist, and it does keep good time. It is a little less accurate than radioactive decay, about twice as sloppy according to Walter Fitch, the man who first drew the cytochrome tree.[19] But this only means that not all nucleotide substitutions are equally likely, which we already know from the redundancy of the genetic code, and despite the play in its mechanism the clock keeps very good time indeed, disagreeing hardly at all with the overall evidence from fossils. Just because one of the disagreements happens to involve our own species — and recall that it is not a clash of facts at all, just an

nd the chimpanzee show evidence of their common brachiating
heritage in the way they use their arms.

Building on the common ape heritage, the gibbon has specialised in efficient movement through the treetops, while the chimpanzee is able to walk on the ground, using the knuckles for additional support.

Humans walk more efficiently than chimpanzees, but given the
opportunity our ape ancestry comes to the fore.

The fossil evidence is sparse indeed. Don Johanson and the fossils from the Afar (top), and (bottom) the *Ramapithecus* scraps at Yale.

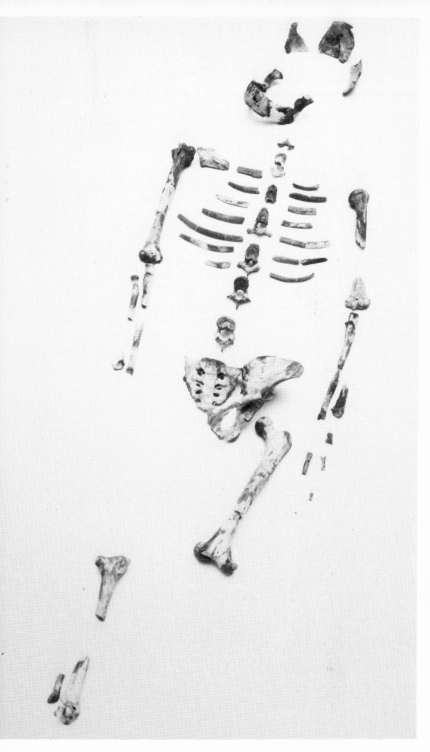

Lucy, the oldest human?

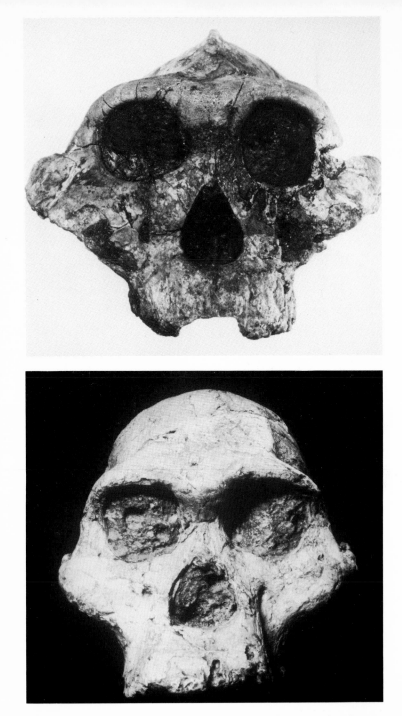
Lucy's descendants, robust (top) and gracile (bottom) australopithecines. Are these the ancestors of gorilla and chimp?

Homo habilis, the first human who is not also an ancestor of the modern African apes.

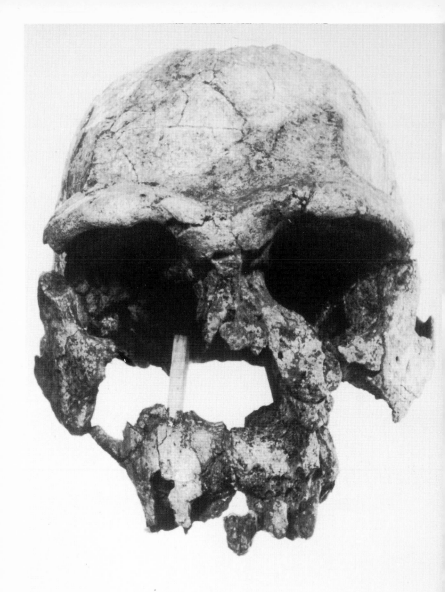

Homo erectus, the species that spread man around the world.

Homo sapiens (left) and the experiment of Neanderthal. Do we still carry Neanderthal genes?

Giraffe and kudu share a waterhole; they also share an evolutionary heritage. Traditionalists are puzzled by the absence of fossils with intermediate necks, but the modern theory allows dramatic changes in a very short time, because small mutations to the genes that control development have a profound effect on final form.

The life cycle of the common frog illustrates the importance of control genes. The embryo, tadpole and adult share 100 per cent of their DNA.

The axolotl, like the frog, goes through a tadpole stage. Unlike the frog, most axolotls never become 'adult', but are able to breed nevertheless. Man might be an ape that breeds while immature.

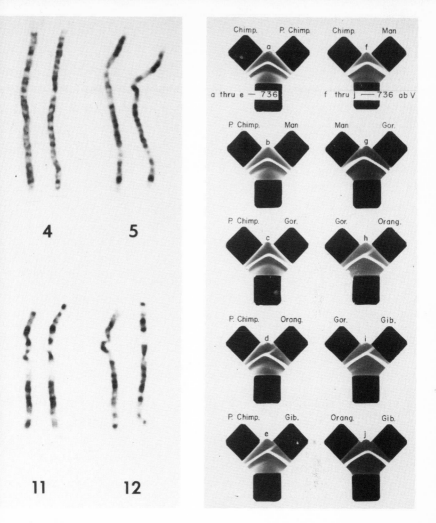

The closeness of man and apes is revealed by a host of techniques. The chromosomes (left) are practically identical – in each pair in this selection the human chromosome is on the left of the picture and the chimpanzee on the right. Immunodiffusion (right) reveals the similarity of the blood proteins. The more symmetrical the pattern, the more closely related are the two species.

Pinnipeds have also been disentangled with molecular methods. The eared seals, as exemplified by the Australian sea-lion (top), and the earless seals, like the nothern elephant seal (middle), share a common marine ancestor. The walrus (bottom) split from the pinniped stock at about the same time that man and the African apes separated.

It all began here. This mouse lemur is very like the primitive primates from which all modern forms are descended.

Human races vary enormously in their physiognomy, but the differences are not adaptive.

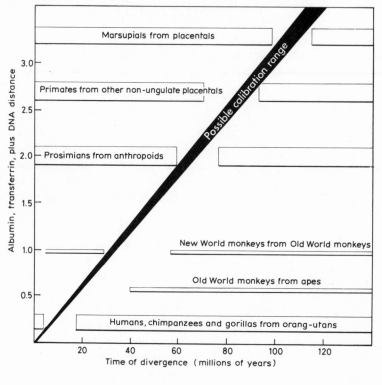

Groups Compared	Albumin	Transferrin	DNA	Time of separation (millions of years)
Man – Chimp – Gorilla	0.12	0.13	0.15	4 – 5
M-C-G – Orang-utan	0.25	0.30	0.33	9 – 11
Apes – Old World Monkeys	0.58	0.53	0.61	20 – 22
Old World – New World Monkeys	1.00	1.00	1.00	35 – 38
Higher Primates – Prosimians	2.10	1.80	2.60	70 – 75

Where fossil evidence is available, it sets upper and lower limits on the dates at which different lineages separated. Any calendar of evolution must pass through these windows; the molecular clock only does so if the man-ape split occurred 4½ million years ago. Horizontal bars represent non-molecular evidence and the diagonal line is the molecular clock.

opinion holding out against the facts — is no good reason to set the clock aside when we consider *Homo sapiens*. If it is good enough for all the other species, it is surely good enough for us. Referring to the many attempts to 'explain' the clock data away by the various objections we have outlined, Vincent Sarich has this to say: 'To stretch a lineage or set of lineages in time is to contract others, which ultimately means that an observed regularity is seen as being a chance combination of grossly irregular rates of change. This is of course possible in the abstract, but it seems to us that to choose such an "explanation" of the hominid data is clearly anthropocentric — an implicit statement that somehow the rules of the scientific game just do not apply to the study of man.'[20]

Man's uncanny resemblance to the African apes, and even more his very recent shared history with them, are established beyond doubt. It would be a good idea now to take stock, and measure the similarity of man and the apes against other relationships in the animal kingdom, for one of our major guiding principles is that in all our investigations of man and his history we should be sure to treat man as if we were Martian zoologists and he just another of the 5,000 mammals alive on the Earth. Within a single breeding population — and for the sake of argument we can regard all people as members of a single population — the differences between two individuals typically amount to some 6 per cent of their genes. The chromosomes of two parents, mixed to form a new individual, probably differ by this much, and this difference is the basis of the enormous variety we see around us in the people of the world. This figure, about 5 or 6 per cent, seems to be typical of widespread vertebrate species like our own. Two seagulls differ by the same amount, as do two mice or two rats. For many invertebrates, the differences can be much greater, and around 15 per cent of their genes will be different in two fruitflies of the same species. The differences between people from ostensibly different races are, incidentally, no greater than the differences between individuals from the same race, which rather gives the lie to anyone who believes that human races are biologically meaningful entities. While the two authors of this book probably differ from one another in about 6 per cent of their genes, neither of us would differ by more than 7 or 8 per cent from any other person on Earth. However, while it happens to be the case that the division of the human species into races is biologically meaningless, the same is not true for other species. An American Indian and an Australian Aborigine will breed as successfully as two Indians or two Aborigines, but a frog from Spain would not do as well with a Rumanian breeding partner as with another Spanish frog. In such cases, where we pick two individuals from different breeding populations, which may well be designated as belonging to different races, the genetic difference between them will be a little greater, perhaps around 8 per cent. Differences are even greater for so-called semi-species, which can interbreed but don't, and for sibling species, which cannot interbreed but are physically almost indistinguishable, like the willow warbler and the chiff-chaff, two little birds that can only really be separated by hearing their different-sounding songs. The biochemical evidence mirrors speciation in progress, with increasing distances between increasingly distinct groupings.

Within this overall context of the differences between different groups, the similarities between man, chimp and gorilla are revelatory. The three are more similar than horse and zebra, dog and fox, and about the same as the two races of the same species of gopher that live on opposite sides of the Colorado river. We feel confident that any biologist who was himself neither an African ape nor a man would have no hesitation in classing the three of us not as distinct species but either as subgroups of the same species or as sibling species. Sibling species, remember, are defined as those which behave as different species and do not interbreed but which are physically so similar that they can only be distinguished from one another with great difficulty — an almost perfect description of the relationship between man, chimp and gorilla, at least at the molecular level.

Nor are the anatomical differences actually all that great, though they seem huge. Carl Linnaeus, the great anatomist and pigeon-holer, put man in a separate category all of his own, though he regretted this in later life. 'I demand of you,' he wrote, 'that you show me a generic character . . . by which to distinguish between Man and Ape. I myself most assuredly know of none. I wish somebody would indicate one to me. But if I had called man an ape, or vice versa, I would have fallen under the ban of all the ecclesiastics. It may be that as a naturalist I ought to have done so.'[21]

We should also take the time to clear up a confusion that may have arisen in your mind. Any two people, we've said, differ by about 6 per cent of their genes. Man and chimp, we also said, differ by less than one per cent of their DNA. How can this be? Surely man and chimp, no matter how close they are, must be further apart than man and man? Of course they are, and the confusion arises because we are measuring the differences in several ways. 'In the first case,' Vincent Sarich explained to us one cold winter afternoon as we sat having tea in London's Knightsbridge, 'the 5 per cent difference refers to whole genes. That is, if I selected twenty genes at random from two individuals, the chances are that one of those genes would be different in the two. But each gene is a string of DNA, a thousand units or more long, and the difference might be in just one of those units. So when people talk about a 5 per cent difference in the genes, they are really talking about a much smaller difference — probably around one tenth of a per cent, taking the silent differences that don't show up into account — in the DNA itself. The figure of one per cent for man and chimp and gorilla relates directly to the DNA, not to the genes, so it is in fact considerably larger than the difference between two members of the same species.'[22] It is all a question of standards, and the confusion is the same as the confusion that might arise if a draftsman mixed inches and centimetres on a blueprint. To make matters worse, there is still another scale of measurement available, rather like furlongs, or cubits. This type of measurement, often spoken of as 'genetic distance', picks on one level of difference as being a standard — a cubit was the difference from Pharoah's elbow to his fingertip, and changed when the Pharoah changed — and then relates everything else to that standard. For example, it would be easy to measure the average immune distance between different species that are also placed into different genera, and arbitrarily state that this level of difference would henceforth be equal to one unit. The average distance between other levels can then be expressed relative to this standard. Like the straightforward immune distance, and the percentage difference in the DNA, this type of measurement shows that the distance between man and the African apes is considerably less than that between animals from two species placed in separate genera, and also less than for typical sibling species. Indeed, there are some single species in which two individuals selected at random are likely to be more different from one another, measured in terms of the sequence of letters making up their genetic code, than a human and a chimp, but although these examples point up the astonishing similarity between man and chimp, we must

(130)

confess that they are pretty rare. By and large, the distance between man and chimp is more than the difference between members of the same species, but not by very much.

All the different methods of measurement have the potential to confuse, but they need not. Just as the draftsman can, to avoid confusion, specify the scale he is using, it should be clear when we are talking about differences in genes, and when we are talking about differences in DNA, the former being about a thousand times more numerous. In any case, one thing is inescapable: regardless of the markings on the ruler, man, chimpanzee and gorilla are far closer together than anyone had previously imagined. Jorge Yunis, the Minneapolis research physician who has done so much to uncover the identity of the great ape chromosomes, speaks for everyone when he expresses his perplexity. 'Such a remarkable degree of similarity,' he points out, 'makes difficult a precise explanation of the large biological differences observed between these two closely related species.'[23]

If the evidence is so good, why has it not been welcomed? To tell the truth, we don't know, but we do know that the popular image of the scientist as a dispassionate seeker after truth could not be further from reality. In the learned treatises about how science is done, there is supposed to be a three stage process. First, the scientist collects some facts. Then he comes up with a hypothesis that might account for those facts. The hypothesis also makes certain predictions about what other facts it should be possible to gather, so the scientist goes off and tries to gather those facts. If he succeeds, the hypothesis stands; it is not proved correct, for you can never prove that a hypothesis is correct, but nor does it have to be discarded. If he fails, and comes up with facts directly contrary to the hypothesis, then the hypothesis is proved wrong (which is perfectly possible) and the conscientious scientist will discard it and try to find a new one that fits the facts better. That's the way it's supposed to be, but in real life things are seldom that simple. Scientists are people, and they have hopes, beliefs and desires just like all people. Sometimes – often – those hopes, beliefs and desires, and fears too, get in the way of the dispassionate alternation of hypothesis and test. So the actual progress of science is not so much a smooth climb to ever higher states of knowledge as a series of conversions as the ruling system of beliefs is overthrown. It has happened before, when Galileo, Copernicus and Kepler overthrew Ptolemy and put the Sun at the centre of the Solar System, relegating the Earth to a mere planet, circling the Sun like all the other planets. It happened when Newton overthrew Kepler, and when Einstein overthrew Newton. Someone may yet be waiting in the wings to replace the system of knowledge that Einstein provided. Ideas do change, and our understanding does grow, but only slowly.

In biology there have not been so many changes. The first was probably also the most important: Charles Darwin's formulation of the laws of evolution by natural selection. Here too, there has been a delay while the establishment assimilated the fresh ideas. Darwin's theory was quickly grasped and appreciated by such men as Thomas Henry Huxley, later to become 'Darwin's bulldog', who had been in on the later stages of Darwin's development of the ideas. (Indeed, Huxley is reported to have said, when told of evolution by natural selection, 'How extremely stupid not to have thought of that'.) But it was also pilloried and vilified by the establishment figures of zoology, and of course the Church. All this is understandable, but perhaps the most fascinating episode in the whole history of evolutionary thought was the lack of any enthusiasm for the original publication of the ideas.

In the high summer of 1858 Darwin had been shocked to receive a letter from Alfred Russel Wallace, a young naturalist out in the Malay archipelago with whom he had been in correspondence. In this letter, Wallace explained to Darwin the essence of a theory of natural selection that was uncannily similar to Darwin's own theory, even down to having been inspired at least partially by Thomas Malthus' essay 'On Population'. Darwin had been working on what he called the species question since 1838, and had written an outline sketch of his theory back in 1842, but he had hesitated to go public with it. Wallace's letter changed all that, and on the advice of his closest friends he agreed to a presentation of the theory to the members of the Linnean Society. Darwin did not attend the meeting, and of course Wallace was in Indonesia, but on 1 July 1858 the Linnean Society heard a paper that was recorded in the leather-bound journal of the society's Proceedings thus: 'On the Tendency of Species to Form Varieties; and on the Perpetuation of Varieties and Species by Natural Means of Selection. By CHARLES DARWIN Esq., F.R.S., F.L.S., & F.G.S., and ALFRED WALLACE Esq. Communicated by Sir CHARLES LYELL F.R.S., F.L.S., and J. D. HOOKER Esq., M.D., V.P.R.S., F.L.S.

&c.' That paper contained all the essentials of the theory that came to be called, even by Wallace, Darwinism. The theory later aroused a huge row, public and vitriolic, and yet the first public statement of those ideas was a non-event that resulted only in a deafening silence. As Darwin recalled in his autobiography some 20 years later, 'Our joint productions excited very little attention, and the only published notice of them which I can remember was by Professor Haughton of Dublin, whose verdict was that all that was new in them was false, and what was true was old'.[24]

But the most outstanding event, the one that really makes us wonder about some of the Victorian men of science, took place eleven months later, on 24 May 1859. The occasion was the annual meeting of the Linnean Society, held every year on that date to commemorate the birthday of Carl Linnaeus, the great Swedish scientist and the man who gave us our present system of taxonomy. In the chair was Thomas Bell, a dentist by training, who as president of the Linnean Society had also been present on the night that the Darwin-Wallace paper had been read. 'The year which has passed,' Bell said in his end-of-year speech, 'has not, indeed, been marked by any of those striking discoveries which at once revolutionise, so to speak, the department of science on which they bear; it is only at remote intervals that we can reasonably expect any sudden and brilliant innovation which shall produce a marked and permanent impress on the character of any branch of knowledge, or confer a lasting and important service on mankind.'[25] He went on in this vein for several more minutes, bemoaning the fact that the previous year had not seen the emergence of an intellect of the stature of Bacon or Newton, Oersted or Wheatstone, Davy or Daguerre. Now it is true that the 1 July meeting of the Linnean Society had been an extra-ordinary meeting — called to elect a successor to Robert Brown, a vice-president of the society who had recently died — rather than one of the normal Thursday evening meetings held every fortnight. And it is also true that six papers, an unusually large number, were read at that meeting. J. W. T. Moody, a librarian who has made a special study of the documents of the time, explains the silence that greeted the Darwin-Wallace paper as follows: 'The fellows [of the Linnean Society] were not so much stunned by new ideas as they were overwhelmed by the amount of information loaded on them at the meeting'.[26] Even so, Thomas Bell had eleven months in which to appraise the Darwin-Wallace contribution to his society's transactions before he gave his presidential address. Yet he ended his review of the past year diplomatically rather than wisely, saying 'I abstain from particularising any of the papers as especially interesting or valuable, as selection would be invidious where all are good, and every student will be able to judge for himself of their respective value and importance'.[27] We, along with history and every other student of biology, have made our judgements; the contributions of Charles Darwin and Alfred Wallace to that summer's evening meeting of the Linnean Society are of the highest value and importance. That Bell not only didn't realise it, but also remained an anti-Darwinian until his death, is further vindication of the unknown biographer who said, of Bell's time as professor of zoology at King's College London, 'in this capacity he made no mark'.[28]

Still, that's all ancient history, and while it affords us a wry chuckle now we have surely learned our lesson. Or have we? There is a curious echo of the reception Darwin's original paper received in the treatment afforded Sarich and Wilson's 1967 paper. No reviewer picked that out as a highlight of the year — but by 1982 it is already clear that future historians of science will be sure to give it prominence. The problem is that although Darwin removed man from centre stage he was still seen by many people as the central player in the drama. Evolution actually removes any possibility of man, one species among the millions, being anything other than just another species. And the molecular clock is the proof of man's ordinariness. True he can ponder his existence, go to the Moon, and kill his own and other species at will, but he cannot open his jaw like a snake, nor soar like an eagle, nor breed like a rabbit. And yet that singular ability to ponder his destiny has led to man setting himself apart from the rest of nature, building, as Stephen Jay Gould puts it, a picket fence around himself. It was surely that feeling, that man must be special, that created a great deal of the hostility to evolution, and still poses problems for the general acceptance of the molecular clock. Darwin was certainly aware of the furore that his theory would create, not because it explained evolution among the inhabitants of the natural world, but because it placed man firmly within, not apart from, that world. In *The Origin of Species* he skirted the issue almost entirely, saying only that 'light will be thrown on the origin of man and his history'.[29] Huxley, five years later, published a brilliant essay on the 'Evidence as to Man's Place in Nature', which used the ideas of natural selection to demonstrate our close affinity with the African apes. And only in 1871 did Darwin finally publish *The Descent of Man,* in which he explained how the theory of natural selection applied to man as surely as to all other species.

Man has always been the major stumbling block in biology. Many scientists were quite prepared to welcome evolution by natural selection for the whole plant kingdom and for all animal species except their own. They simply could not accept the ultimate uniformity, that what applied throughout time to all other forms of life also applies to our own species. Wallace himself suffered this blind spot, and while he was quite content that man's physical characteristics had been shaped by the selective hand of evolution, baulked when it came to mind. This noble creation, Wallace believed, was given to man by a higher form, a deity. Today this feeling lingers on, not in a denial of evolution but in a desire to place man as far back in time as possible, safely distancing him from the other animals. The molecular evidence threatens that safety by making it clear that we are not 20 million years away from chimpanzees but less than 5 million years. And the response to this information is the same as the response to threatening information has always been: deny it.

We like to think that Darwin would be more than pleased at the way things have gone, and especially at the way so many of his shrewd guesses have been proved correct. He guessed that life would have arisen in a warm pond, and he was almost certainly right. He guessed that man evolved in Africa, and again he was right. He said, on the basis of anatomical evidence, that man's closest relations were the chimpanzee and gorilla, and he was right, but how right he never knew. The molecules tell us the three of us are more than 99 per cent identical, and that we shared an ancestor with the African apes less than 5 million years ago; we also like to think that Darwin would have welcomed this information. Today, however, instead of welcoming this information, many experts still deny its relevance. They accept it right enough for other species in other groups, but when it comes to primates, and in particular *Homo sapiens*, they turn away from the molecular evidence. Darwin would never have countenanced this egocentricity. Admittedly he dodged the issue of man in *The Origin,* but this was a political decision, designed to avert the Church's wrath, something it failed to do very well, for even the Church could see that man was included within

(135)

Darwin's scheme of things. The Church, too, is one of the great bastions of anthropocentrism; did not God, after all, make man, and only man, in His image, and did He not give man dominion over all His other creations? One can fault *The Origin* for sidestepping the evolution of *Homo sapiens,* but one cannot fault Darwin's good sense in doing so, nor find in this any reason to doubt his own belief in the evolution of man.

The Descent of Man sets out his ideas and argues forcefully for a common relationship between apes and men, but it does more than that, for at the very start of the book, to open the first chapter, Darwin sets out some of the criteria that we need to adopt if we are to decide whether 'man is the modified descendant of some pre-existing form'. He says that we must look to see whether man varies, and if so whether those variations are transmitted 'in accordance with the laws which prevail with lower animals'. He tells us to investigate the causes of the variation, to check that they are 'governed by the same general laws, as in the case of other animals'. In all the criteria that follow, Darwin is at pains to point out that man must be seen to conform, in this way, with other animals, and his book consists in large part of the evidence that man is indeed just like all the other animals. From there it is a simple step to conclude that man did indeed descend from some ancestor.

We have taken Darwin's words as our motto because they are as timely today as they were in 1871. They tell us how to approach the study of man, and they have guided us in our interpretation of the molecular evidence. There can be little doubt that we are not only incredibly similar to the African apes, but also that we are very recently emerged from the common ancestor we share with them. Why should this be so very disturbing? Why can people not accept that this information does nothing to destroy man's supposed dignity, that it does not threaten his position as most powerful animal? We do not know, but we do think it important to learn from the lesson of Thomas Bell and recognise a revolution when we see one. The palaeontologists will eventually join us.

Remembering Darwin's advice, and using the evidence of the molecules to supplement the meagre fossil record, we now have, in more detail than ever before, a picture of the evolutionary tree that has at the tips of its branches the species alive today. In almost every respect this agrees with the tree provided by fossil evidence and comparative anatomy, in itself striking proof of the reality of evolution. We also have something that the fossils could never really provide us with, and that is a set of good dates for the forks of the tree's branches. About 90 million years ago a primitive, insect-eating animal very like a tree shrew split off from the other mammals to found the primate line. Fifteen million years later, 75 million years ago, that line split five ways in the major radiation of the primates, with another split 50 million years ago to divide the lorises from the lemurs. On our line, nothing much happened until about 35 million years ago, when the New World monkeys broke away — literally — and drifted off on the South American continental plate during the break-up of Gondwanaland. Our line continued for another 15 million years, but about 20 million years ago there was another major split, between the Old World monkeys and the hominoids — apes and man. Keeping, selfishly, on our own line and ignoring the fascinating changes taking place among the other primate groups, the next split of interest is about 12 million years ago, when the gibbons take their leave of us, followed 4 million years later by the orang-utans. The split of most interest to us, when man, the chimpanzee and the gorilla separated from each other, took place about 4½ million years ago. Around 2 million years ago, as man was beginning his meteoric shift to become the cultural animal, the chimp split into the two present-day species, the smaller, forest-dwelling pigmy chimpanzee (*Pan paniscus*) and the larger inhabitant of the wooded savannah (*Pan troglodytes*), and the gorilla divided into the lowland subspecies of West Africa and the mountain gorilla of the East. There may well have been a split in our own line at the same time, but we are the only survivors.

Those, then, are the branches of our tree and the dates at which they forked, an outstanding achievement of modern biological science. But there is more that we can add to the story, for now that we know the path taken by our ancestors as they evolved into us, we can begin to examine the intriguing mystery of why the particular changes occurred when they did, and how a bunch of genes made the astounding transition from monkeys to men.

Six

MONKEYS TO MEN

The very first primate, 90 million years ago, was much like the present-day tree shrew, a small insect-eater that resembles a large mouse. Today one species of primate, man, numbers more than 4½ billion members and can be found almost everywhere on the surface of planet Earth. The route taken by man's line from the first primate to the present day we know, more or less, but there are details to be discovered. What caused the tree shrew to stay as it was, while the other primates changed, some of them dramatically? Why did some primates first change their way of life in the trees and then abandon the trees? What were the selective forces that favoured walking on two legs and developing a large brain? These are the questions that must be answered if we want to understand how our ancestors were transformed from monkeys to men, and to do so we will need far more from the ancient rocks than the bones they contain.

Evolution comprises the relentless change of living things that enables them to take better advantage of their surroundings. Whenever the surroundings change, the life-forms must perforce change too, and in the history of all species the major changes are associated with major ecological changes. Man is no exception. At the time the primates emerged from the mammalian stock, the Earth was covered with a lush green carpet of forest. By the time man emerged from the primate stock, the forests had given way to open savannah, and the transitions along the way shaped the primate line. We will come on to the reasons for these changes later (in Chapter Eight). What matters here is that the early primates were not so much tree-dwellers as inhabitants of the undergrowth, scurrying about in the bushes looking for insects and suchlike to eat. The death of the dinosaurs 65 million years ago opened the way for all sorts of animals to move into the niches vacated by the dinosaurs, and probably at the same time removed many of the predators that had hitherto kept the mammals small and secretive. Out into the day came the primates, and up into the trees they ventured.

This is perhaps the place to reiterate that although we may talk of evolution in purposive terms, suggesting for example that animals evolved an opposable thumb in order to grasp branches or manipulate their food, this is not the way evolution really works. The mechanics of sexual reproduction ensure that variations are thrown up in every generation, and the inevitable struggle to reproduce selects among the variants those best suited to take advantage of the environment. To return momentarily to the cumbersome language we ought to use, those animals that had opposable thumbs were better able to manipulate their food, were more successful as a result, and came to dominate the population. Later on, when thumbs became a hindrance for certain primates, those animals with a smaller, less obtrusive thumb were preferred and the advantages of superior manipulation, no longer so valuable, were traded for improved locomotion. Evolution has no purpose, no sense of direction, but it is clumsy to use the strictly correct form of words all the time so we prefer to be more direct; evolution isn't really like that, but so long as this is clear the actual form of words we use won't matter too much. That said, we can return to the evolving primates.

The climate all those millions of years ago was warm and equable, almost tropical, and in the north of Laurasia, where London is today, palms and cycads flourished, as did magnolias, cinnamon trees, and even figs. Much of the land, it seems, was covered with rather dense forest, bordering here and there on swampy pools, and it was in these northern areas that the mammals began their upsurge. One of the first primates that we have any evidence of is called *Purgatorius*, and it is very like a tree shrew. The fossils of *Purgatorius* were found in North America, which today has no native primates other than man; this surprise is explained, like so many mysteries of times past, by the inevitable drift of the land masses around the globe. Sixty million years ago North America and Europe were part of one land mass, and it was on that land mass that the primates evolved, so it is no real surprise to find early primate fossils in America. Also evolving at the same time was the rodent family. Similar in many respects to the early primates, the rodents had one big advantage: their teeth never stopped growing. This meant that they could exploit tough foods — seeds, nuts and so on — with much more efficiency than the primates, whose teeth did not grow continuously and were worn down by this diet. Throughout much of their range the more specialised primates, those that had given up insects for a diet of grains and plant material, came under threat from the rodents, and many of them succumbed. Only the more generalised primates survived, and they did so in Africa, where for some unknown reason they were safe from the rodent menace.

Forty million years ago the massive land mass of Gondwanaland began its final break-up, and South America started its stately progression away from Africa, the South Atlantic ocean widening by a couple of centimetres each year. One group of primates was trapped on the drifting continent, and eventually became the New World monkeys, while a separate group was left behind in Africa. (It's all relative, of course. The monkeys in Africa saw their relatives drifting off, and geologists follow the same convention of regarding Africa as the fixed point and relating all other plate movements to it.) The two groups evolved very much along parallel paths, at least up to a point, and the differences between New and Old World monkeys are neither plentiful nor obvious. The nostrils of a New World monkey are further apart and face sideways, while those of an Old World monkey are closer together and face downwards, and some New World monkeys have evolved a sensitive, highly co-ordinated tail that they use as a fifth limb, but otherwise the two groups are very similar. The changes that they both show from the more primitive earlier primates are largely the result of two important evolutionary shifts, a move into the trees, and a switch from being active at night to exploiting the hours of daylight.

Some of the early rodents were probably fairly agile in the trees, but in Africa and South America, where the primates underwent their major changes, few animals were taking advantage of a life in the trees. But the advantages there were many. For a primate that could make use of it, there was a plentiful supply of food, as the plants were also evolving rapidly at this time. It may surprise many readers to learn that flowers, and with them fruits, began to appear only towards the end of the age of the dinosaurs; the tree-dwellers would surely have taken advantage of this relatively new food source. Leaves, too, are nourishing, especially young ones, and in the tropical forests of the time there would have been a year-round supply of young shoots and fruits. Predators were in all probability scarce, especially at the beginning. But the requirements of this sort of lifestyle were different from those of a nocturnal insect-eater scurrying around in the dark on the forest floor. Eyes, for one thing, would have to change, for the sensitive eyes of a night creature would not be so useful in the daytime. One of the major changes of the time was probably the development of colour vision; among the green leaves brightly coloured fruits would be highly visible, but only to a creature with eyes that were sensitive to different wavelengths. The eyes of a nocturnal animal are sensitive, in that they respond to faint illumination, but the picture they provide is composed only of shades of grey. Daytime eyes are not so sensitive to faint light — they have no need to be — but they provide a great deal of extra information about colours. The very fact that ripe fruits are, by and large, conspicuous to our eyes is testimony to the fact that colour vision and coloured fruits probably evolved in tandem. It is advantageous for the fruit-eater to be able to find and eat the fruit; this is also advantageous for the fruit plant since that is how its seeds get scattered. But the plant does not want its fruits eaten before the seeds are ripe, that would be a waste of the resources it has put into the fruit, so unripe fruits are not only green and hard to see, but also protected with a number of poisonous chemicals that disappear when the fruit is ripe, ready to eat, and conspicuous. Evolution has selected fruits that are brightly coloured and appeal to us, at the same time as selecting eyes that can see ripe fruits and hands that can pick them.

The position of the eyes would have to change too. A small animal of the night uses its eyes less to find food than to detect predators; food, and direction too, are taken care of by the sense of smell, but long-distance detection of danger depends on acute hearing and sensitive eyes. Most nocturnal animals have quite large eyes, which also cover a wide field of view. With the exception of predators like the owl, the eyes are placed on either side of the head, where they can scan a major part of the surroundings. Primates, by contrast, have both eyes on the front of the face, where they face forward. The field of view is restricted, but both eyes cover the same area. This is called binocular vision, and we tend to take it very much for granted. But binocular vision was a major evolutionary advance: it enables us — and the owl — to see depth clearly. Because they are not in exactly the same place, each eye has a slightly different view of the world. The brain takes the two slightly different pictures provided by the eyes and performs a sophisticated analysis that yields, from two flat pictures, a model of the real world with depth. Just try picking up a pin in a pair of tweezers with one eye closed; it's pretty difficult, because the depth information that binocular vision normally supplies is missing. In the trees, binocular vision and an accurate sense of depth was obviously a boon, for it enabled the primates to judge accurately the distances between branches, the spacing of the things round them, and the location of important objects. That all primates, except the tree shrews, have forward facing, binocular eyes, shows that this was one of the most useful of the primate adaptations.

As the eyes came to the fore, so the nose receded. Smell is still important to the primates, even to man, much of whose social life is governed, albeit unconsciously, by smell, but it is nothing like as important to the primates as it is to the rodents, say, or dogs. An acute sense of smell is best served by a wet nose, the better to dissolve airborne molecules, and a long muzzle that contains plenty of room for the special sensitive skin that responds to those molecules. Primates have a much shortened muzzle, and a dry nose.

The final major change to primate anatomy concerned their paws. Flat paws are fine for padding along the ground, and even for life in the trees. Many arboreal species have paws and claws that they use to provide a secure hold on branches, but primates have very different ends to their limbs. The first digit, on hands and feet, is constructed in such a way that it can be opposed to the other four digits; this arrangement allows its possessor to hold on to things very effectively, which the primates do as they climb among branches (except, of course, in man, who lost the ability to grip effectively with his toes as part of the price he paid for upright walking). The hand and thumb not only provide more security in the trees for a larger animal, they also allow that animal to manipulate its environment much more effectively than do paws. It's true that no other primate has the precision of man's grip, but they do have a manipulative ability that far exceeds that of most other animals, and that ability also enables them to take advantage of many different food sources. And once you have a hand that can grip, claws are no longer so necessary and may be a positive hindrance to the smooth working of the hand. So in most primates we find that the claws are replaced with nails, flat sheets rather than slender points. Nobody is quite sure what nails are for; perhaps they are non-functional rudiments, proof that evolution cannot totally ditch some structure just because it no longer serves a useful purpose. Or perhaps they provide protection for the sensitive finger tips. They are undoubtedly useful for scratching, but other animals manage without so that cannot be the whole story. Some primates have found a new use for their fingernails and have grown them into pseudo–claws, which they use for winkling insects out of cracks in tree bark, but by and large these particular adaptations are a bit of a mystery.

The basic primate, then, some 40 million years ago, had forward facing eyes and a reduced muzzle. Its hands and feet each had an opposable thumb, and it made a living in the trees, eating fruits, leaves, the occasional insect, and whatever else it could find. Predators were probably few, and the living was probably easy. In South America it continued that way, and many experts consider the New World monkeys to be more primitive, more like their ancestors, than the Old World monkeys, for while the primates in South America continued much as they were during the next 35 million years, the Old World was the scene of further changes and developments. Climatic and environmental change came first, with the primate line — and others — splitting and evolving to produce new species to take advantage of the changed conditions.

During the Oligocene epoch, between 38 and 22 million years ago, the climate and vegetation of the Old World was still much as it had been, lush forests surrounding swampy pools, but it was already beginning to change. The forests were becoming more open and the land was becoming drier. It was at this time that the monkeys in Africa split again, into the ape line and the true monkey line, but in South America it continued that way, and many experts consider evolved there. As we saw in Chapter Three, the earliest ape is an animal called *Aegyptopithecus*, who lived about 28 million years ago in the forests that grew in the Fayum depression of Egypt. *Aegyptopithecus* differs from modern monkeys mostly in the structure and shape of its teeth, though in behaviour it was probably very like a modern baboon. By this time the primates had adapted to their own niche in the trees and no longer faced any threat from the rodents. They were able to spread out again from Africa, into Europe and Asia. The next important apes are the dryopithecines, now known from a variety of sites but originally discovered in France in 1858. Dryopithecines bring us from the Oligocene 28 million years ago well into the Miocene, where they inhabited heavily forested areas of Africa and Europe less than 20 million years ago. The tropical forests now found in Central Africa then extended much further eastward, and covered what is now Kenya too, but East Africa is not the major area of concern to us now, for we are approaching the time when the ancient Old World apes gave rise to the modern lines that are of so much interest, and the major shift here took place not in Africa but in Asia.

How do we know? In truth, we don't, but it is a very plausible guess, convincingly argued by Vincent Sarich. The earliest apes differed from monkeys mostly in their teeth, but they did not possess the unique adaptation that characterises the living apes, the ability to brachiate, to swing through the trees. The gibbons, the siamang and the orang-utan are all brachiators par excellence, and they live in the forests of Asia. The chimpanzee, gorilla and man are also brachiators, though they do so less often and with less agility than the Asian apes. Also, they are found in Africa, not in Asia. The conclusion is that the brachiating adaptation arose in Asia, where it continues to this day, and that the changes that give rise to the African apes took place, not surprisingly, in Africa. We have already suggested some of the reasons why brachiating was a successful step for those early apes; it enabled them to exploit an abundant food source and provided the extra mobility needed to get around safely in the forest, but the key point is that the earliest apes were not brachiators.

There are a few ape fossils from the early parts of the Miocene 20 million years ago, but unfortunately most of those are teeth and jaws, neither of which can tell us much about locomotion. Nevertheless, one ape of the time, *Pliopithecus*, is known in some detail, and has a long tail. That in itself argues that it was not a brachiator, for a tail, while very useful as a balancing pole to an animal running above a branch and to help orient the body for a correct feet-first landing above a support, only gets in the way of a body swinging below a branch. No brachiator has a tail, not only because no brachiator needs one, but also because a tail is a positive hindrance to the brachiating way of life. But in the middle of the Miocene, some 15 million years ago, the dryopithecine apes were going through a major period of transition. One fossil of the time is *Ramapithecus*, but the chances of this being on the direct line to man are slim indeed. *Ramapithecus* was one of the dryopithecine apes that shows in its teeth (almost all the remains that we have) some changes that indicate it was eating a rough diet. We cannot be certain that it was not a brachiator because the relevant skeleton is missing, but the evidence of diet does not fit well with an animal making its living among the trees. *Ramapithecus* was probably an evolutionary response to the shrinking of the forests that took place in the mid-Miocene; it was larger than most of the earlier dryopithecines, and probably dwelt in the mixed woodland at the edge of the forest, where it ate whatever it could find. Although much younger than *Dryopithecus*, even at 14 million years old, *Ramapithecus* is certainly not the earliest hominid, for that wasn't to appear for almost another 10 million years. Indeed, *Ramapithecus* cannot even be placed on the direct line to that ancestor, because our line must, at some stage, have developed as a brachiator.

The apes of the mid-Miocene were faced by a number of evolutionary challenges. In some areas the forests were beginning to shrink, giving way to more open mixed woodland at their edges. Elsewhere the trees provided a bountiful supply of food for the taking, but only to an animal that could get out to the tips of the branches where the trees sequestered their fruits. Some dryopithecines — they were a numerous group, with many species all over Europe, Africa and Asia — moved into the woodland areas and eventually gave rise not only to *Ramapithecus* but also to its cousin ramapithecines, *Sivapithecus* and *Gigantopithecus*. Others probably adapted in different ways to exploit the available niches. We only have a few fossils, and it is more than likely that these do not represent all of the variations on the ape theme between 14 and 4 million years ago. One group — a genuine missing link — changed so as to become expert brachiators. Fifteen million years ago there is no sign of a brachiating ape; by 12 million years ago, the date provided by the molecules for the separation of the gibbons and siamang, brachiation is fully developed. We don't know why the brachiating apes developed as they did, though we suspect that it had something to do with the availability of food, and neither do we know why, of all the dryopithecine lines, only the brachiators survived, but that is what happened. The ramapithecines, the only other line we know anything about, were mostly extinct by about 7 million years ago, although *Gigantopithecus* hung on until barely a million years ago; all the other dryopithecines simply vanished. The world is full of monkeys, but there are very few apes left, and those that remain are all descendants of a single line that learned to brachiate.

By 12 million years ago the original apes have split into at least two lines, the ramapithecines and the ancestors of the living apes. At that time a further split in our line took place when the gibbons diverged from the other apes. The so-called lesser apes — several gibbon species and the siamang — have taken brachiation to its extreme as a way of life. Smaller than the other apes, and with very long arms, they hurtle through the tree-tops with an athletic ease that is nothing short of breathtaking. The gibbons have developed their arm-swinging abilities to a fine pitch, making subtle changes to their anatomy that provide maximal efficiency. The hands have become elongated, especially the fingers, to provide a hook that is used not so much to grasp branches as to hang from them. The thumb, in the process, has remained small and out of the way, and is no longer much use for holding on to objects. The legs are short, and much reduced, and the gibbons are, quite simply, consumate brachiators. When moving at speed through the canopy, with legs tucked up into a ball beneath the body, a gibbon almost seems to fly, barely touching each branch before it hurtles off to the next slender support.

The line that the gibbons abandoned − our line − went in another direction, becoming larger and not so suited to fast brachiation. The orang-utan is no less of an athlete than the gibbon, but its speciality is slow gymnastics rather than racing brachiation. The hip joint of the orang-utan is almost as flexible as the shoulder joint, and the feet as manipulative as the hands. The result, as Don Johanson describes it, is a four-armed orange spider, suspended among the branches and using whichever limb is most convenient to move, scratch, or pick fruit. Orangs spend almost all their time in the trees, but unlike gibbons they move at a slow, measured pace. They are large, but despite their bulk are able to feed out at the tips of the branches, and they do occasionally come down to earth to cross a clearing or get to a tree that is otherwise inaccessible. When they do they may walk on two legs but most often use all four. According to the molecules, the orangs split from the African apes about 8 million years ago. We have no fossils of these brachiating apes, and in fact the next good fossil we have is also the first hominid: Lucy, 3¾ million years old, walked fully upright. The chimp and gorilla, separated from our line a few hundred thousand years previously, are basically quadrupedal knuckle-walkers. Somehow, between 8 and 4 million years ago, the arboreal brachiating apes came down to the ground and migrated from Asia to Africa.

Early in the Miocene, 25 million years ago, the forest carpet extended right across the land masses of Africa, Europe and Asia, but over the next few millions of years, because of climatic changes induced by continental drift (see Chapter Eight), the forests began to shrink, so that by 8 million years ago we can be almost sure that there was no longer any forest connection between Asia and Africa. This immediately poses a problem for our story of ape evolution, and suggests the answer. If there was no forest corridor between the two continents, how did the apes get from Asia to Africa? Get there they most certainly did, as we know from the fossils, and if there wasn't a forest to move through then they must have come overland. But the line that we are descended from had evolved for a life in the trees, so how did our ancestors manage the long journey on the ground?

The answer is that they walked, not on two legs but on all fours, and that it was the previous generations in the trees that, para-doxically, allowed them to make the trip. Brachiating changes a lot more than the structure of the arms and torso. It also affects the lower part of the body and prepares it for walking. The reason is that a brachiating animal spends a great deal of time hanging below, or sitting above, branches. In this position the trunk is held vertically, as distinct from the horizontal trunk of the quadrupedal monkeys. The baboon, walking across open ground, does so on all fours, its body parallel to the ground and its soles and palms flat, but the ape, even when on all fours, holds its trunk much more upright. These are changes that began with a brachiating life in the trees, and they enabled the late Miocene apes to walk on the ground every bit as effectively as modern chimps and gorillas do. The long arms of a brachiator can also be put to good effect on the ground, providing support for the upper body, while the legs, having rotated back-wards compared to the legs of a monkey, are very efficient too. The hands and feet of brachiators are less alike than the hands and feet of quadrupedal primates, because (except in the orang-utan who followed a different path of specialisation) they have become specialised for different things, the hands for grasping and the feet for support; but this, and the changes in the wrist that accompanied the development of arm-swinging, mean that the apes cannot use their hands simply as front feet. If they tried to walk on their palms they would push the bones of the lower arm right through the wrist. Instead, apes use their hands curled under, taking the weight on their knuckles and providing a name — knuckle-walking — for their peculiar way of movement.

Knuckle-walking, though you would not think so to read some texts, is neither inefficient nor degraded; it is a very good way of getting around on the ground, especially for a long-armed brachiator recently come down from the trees, and we can be sure that between, say, 8 and 6 million years ago our ancestors made the trek from Asia to Africa in just such a manner. We are not, of course, suggesting that one day a band of large brachiators came down from the trees and headed west until they reached the forests of Africa. More likely a species that was already exploiting the woodland edge of the forest simply moved further afield with each generation, just as 5 million years later their descendants moved out again from Africa to conquer the globe. Whatever it was that actually happened, we can assume that by 5 million years ago a terrestrial knuckle-walker, descended from the brachiating apes, was safely ensconced in Africa. And this is when, from our point of view, the story begins to get very interesting indeed.

We should stress again that the story we have just told differs radically from the accepted scenario as outlined in the textbooks and indeed all other popular accounts. We say that *Ramapithecus* is a dead end, merely the second most successful (as far as we know) of the many adaptations of the dryopithecine apes. Palaeontologists are beginning to shift their ground very slightly, but most still see *Ramapithecus* as an early hominid. We say that the brachiating adaptation arose in one group of dryopithecines about 15 million years ago and that this was when the whole modern ape line became distinct. The palaeontologists don't care much for brachiation anyway, but generally suppose that the significant development of the ape line had already occurred some time before *Aegyptopithecus*, 28 million years ago. We say that the brachiators gave rise to the gibbon and the orang-utan 12 and 8 million years ago, dates based on hard molecular evidence. Palaeontologists say that all the modern apes, including not only the gibbons and orang-utan but also the chimp and gorilla, separated from us about 20 million years ago — although they have no evidence at all to support this claim. Finally, and this is the stage we have yet to deal with, we believe that the evidence is overwhelming that the terrestrial knuckle-walker that trekked from Asia to Africa split three ways just over 4 million years ago, to produce the chimp, the gorilla and ourselves. Once we pass that threshold, when one ape took the terrestrial route to its end while the other two returned to the forests and a more mixed way of life, we are again synchronised with the palaeontologists, and although we may not agree with all of them we are, at least, all talking about the same evidence. Before 4 million years ago, the molecular anthropologists are the only ones with any evidence at all.

The story we have told, of a mid-Miocene radiation of the dryopithecines, followed by the extinction of all lines except a bunch of brachiators, and then the split of that line and subsequent walk to Africa, is not contradicted by a single fossil bone. It is supported not only by anatomy, but by a massive, coherent body of molecular data. The changes that have taken place in the primate line have been many, and the entire collection of fossils is still so sparse that we cannot be sure exactly how those changes took place. We think we know what happened, and why, but we cannot be at all certain. For a long time people thought that human evolution was first a swelling of the brain and then a move to upright walking, and Piltdown man fooled the establishment because it fitted those preconceptions so neatly. We now know, thanks to Don Johanson's finds in Ethiopia and Mary Leakey's footsteps at Laetoli, that a fully upright stance preceded any significant growth in the brain. Some of the ideas that we have advanced in our version of the story leading up to the first hominids might well be like the Piltdown prejudice, sound given the present state of our knowledge but later, perhaps, to be proved wrong. But it does seem important to stress again that this story does at least take heed of all the information available and does not demote any branch of science at the expense of the others. Where there are fossils, we use fossils, but where molecules provide the only signposts, we use them.

The story of the Miocene radiation of the primate group may seem appallingly sketchy, and indeed it is. The problem is that there simply isn't very much evidence on which to base the story. All the known remains of hominid ancestors would fit on a dining table; all the primate ancestors might need a couple of tables, but they still aren't very plentiful, and the tantalising gaps remain. What was going on between 20 and 15 million years ago, when the dryopithecines came to the fore? Even more important, what were the apes doing in the great fossil gap between 7 and 4 million years ago? That was when some of the most crucial steps were taken, and yet the entire primate record of the period would rattle around disconsolately even in a small shoe box. There is less than a handful of fossil fragments from this era.

The hominids of less than 4 million years ago tell us that we're on the right track, and we wait expectantly for the fossil-hunters to push our horizons further back. As we write, Richard Leakey is about to begin a season exploring the sediments on the west side of

Lake Turkana, where he hopes to find support for his belief in a very early origin for our own genus, *Homo*. Don Johanson continues to hope, and plan, for a return to Ethiopia in less troubled times, and when he gets there, he says, he will find 'something between ramapithecid apes and Lucy at around six million' years ago.[1] We will be most surprised if he does, for we do not believe that there is any link between *Ramapithecus* and the hominid line; however, if the history of palaeontology is any guide we won't be at all surprised if he finds something that is interpreted, initially at least, as intermediate between *Ramapithecus* and Lucy. And excavations in Pakistan continue to throw up interesting fossils from the mid-Miocene. Doubtless the palaeontologists will eventually find what they are looking for, and perhaps when they do we will have to revise our story. Alternatively, and we feel this is more likely, someone new and uncommitted will come along and find fossils that support the molecular story, and then, perhaps, the palaeontologists will change their story. Until that day, unfortunately, we will simply be in the dark about those tantalising years during which a knuckle-walker gave rise to a strider. We can, however, pick up the story shortly after the upright ape appeared — but to do so we must travel via a remote mountain lake in Mexico.

Seven

THE APE'S BRAINCHILD

Consider the strange life of the common frog. From a single fertilised egg a new frog will grow, but not in one smooth process. First the single cell of the egg divides to form a bundle of cells, which gradually multiply and reorganise themselves via a number of stages into something that looks rather like a little fish. This something is, of course, a tadpole; it has no limbs but uses its tail to swim and gets oxygen from the water through its feathery external gills. As it grows older the tadpole slowly changes. It grows a pair of hind legs that it uses to help with swimming, and it also develops a pair of front limbs. The feathery gills disappear to be replaced by fully formed lungs that allow the tadpole to breathe air and escape from the water. Eventually the tail too vanishes and the distinctly fishy tadpole has metamorphosed into an amphibian frog, at home on land but returning to its origin, the water, to breed.

Why do we tell this story which seemingly has no connection either with the evolution of man or the mountain lakes of Mexico? Because it illustrates a fundamental principle of evolution: there is very little room for change, only for modification. The ancestor of the frog was a fish, a type of lungfish that had special pouches in its throat that it could use as rudimentary lungs. This animal could escape drought — or lack of oxygen in the swampy waters in which it usually lived — by using its simple lungs to breathe air, and it could crawl from one pond to another in search of water. Those with stronger limbs and more efficient lungs survived and gave rise to the many thousands of frogs, and other amphibians, alive today. But they didn't entirely free themselves from their dependence on water, for the eggs of an amphibian will develop only in water, even if that water is just a few drops trapped at the base of a tropical leaf. Eventually, of course, some of the primitive amphibians did overcome this enormous hurdle and developed the possessions — notably a waterproof skin and an egg with a protective shell — that made them reptiles, and even less dependent on their environment. Those possessions, like all the other attributes of an animal, were enabled by changes to the DNA, but at every stage new blueprints in the

DNA are laid over existing plans, so that rather like a medieval parchment that has been used and re-used many times, there are traces of the old document in the new. The frog baby, the tadpole that emerges from an egg, is effectively a fish, the ancestral form of the amphibians. It cannot reproduce, for that task is left to the adult, air-breathing land–dwelling frog, but the existence of the tadpole is a continuing reminder of the frog's ancestry. The DNA that instructs a cell to build a frog no doubt also contains the information needed to build some sort of fish, could we but decode it. All the genetic instructions for today's living species are built upon the modified instructions for ancestral species.

Most frogs are perfectly ordinary amphibians, respectable and conformist in every way. There are exceptions, of course, whose eggs develop not in free water but in specially modified organs; the midwife toad carries his eggs on his back, and there are some species of frog that swallow their eggs and brood them in a special pouch off the stomach, keeping them safely inside all through the long process of development until the day tiny frogs hop out of their parent's mouth. But in general amphibians lay their eggs in water, where they hatch into tadpoles that eventually metamorphose into the usual adult form. There are, however, some amphibians that have added a strange twist to their life cycle and among these are the beasts that take us to Mexico.

The axolotl is an odd–looking creature, rather like a giant tadpole but with four small limbs, a large propulsive tail and the finely-divided gills needed for life underwater. But while a tadpole is an immature frog and cannot breed the axolotl is perfectly capable of making eggs or sperm and reproducing; in form it is immature but in function it is adult. Sometimes, though, an odd thing happens to the axolotl. Its gills begin to shrink and, finally, vanish. It devlops lungs and sturdier legs and eventually crawls out of the lake and onto land, looking for all the world like a large salamander. Indeed it *is* now a salamander and the salamander itself can also breed, though like all amphibians it must return to water to do so and to lay its eggs. Bizarre though it may seem, the axolotl, immature form of the salamander, has a DNA blueprint changed through mutation to enable it to reproduce itself while still in the juvenile form. This process is called neoteny, which means literally stretching out, or holding on to, youth. Some species of axolotl seem to change spontaneously into salamanders, especially when the pond they are in dries out. Others never seem to make the change in the wild, but can be persuaded to do so in the laboratory by adding a minute quantity of iodine to the water in their tanks.

This is the clue to the evolutionary advantage that neoteny gave to the ancestors of modern axolotls. Iodine is an essential constituent of the hormone that triggers metamorphosis into the adult state, but some of the Mexican lakes inhabitated by axolotls contain scarcely a trace of this vital element. Without the iodine there could be no metamorphosis: any salamanders unlucky enough to lay their eggs in those lakes would have plenty of viable, tadpole-like offspring develop from the eggs but no further reproductive success, because the tadpoles would never become adult and so would never breed. A rare mutation — perhaps only one clutch of eggs long ago — producing a few tadpoles that could accelerate the development of their sexual organs and hence breed while still not metamorphosed, would surely have been an instant success. Salamanders that could breed as tadpoles would quickly have spread through all the iodine-deficient lakes.

What relevance has the bizarre tale of the axolotl to the story of human evolution? A great deal, surprisingly, for both neoteny and water have been claimed as major factors in the evolutionary developments that separated modern man from the African apes. Indeed, looking further back into the past, many experts believe that neoteny was crucially important in creating not just man but the whole group of animals with stiffening rods down their backs, the chordates.

In most chordates the stiffener is made of bone — a backbone — and this group is known as the vertebrates and includes ourselves. Some chordates, however, have a simpler stiffening rod made of cartilage, and the vertebrates must have evolved from a simple primitive chordate with just such a cartilage rod. But where did this simpler form itself come from? There is no spineless animal that looks sufficiently like a chordate to be regarded as a reasonable candidate for the ancestral form. Or rather, to be specific, there is no *adult* spineless animal that fits the bill. There is a creature, the sea-urchin, which looks nothing like a chordate but which shares more subtle biological properties than appearance with the family of backboned creatures, and there are great similarities, both in appearance and at a deeper level, between the sea-urchin *larva* and some of the simpler chordates. This evidence is strong enough to convince many evolutionary biologists that the whole chordate family has evolved from one kind of sea-urchin larva that held on to youth and learned, like the axolotl, to reproduce neotenously.

All this occurred more than 400 million years ago, at the start of the era that geologists call the Silurian, long before the distinction between man and chimp, but the example highlights a crucial and interesting aspect of neoteny: it allows a species to spring itself from the trap of specialisation. This sounds melodramatic, so we should explain what we mean. Adult animals tend to be well designed — well adapted — for a particular way of life, but juveniles and infants, still in the process of growing and developing, are generally less well fitted to a particular adult niche. Furthermore, remember that all species develop from almost identical balls of cells, with their special features becoming apparent only later in life after development through many stages. Now imagine what happens if some new environmental opportunity should present itself. The adult — a sea-urchin, for example — specialised as it is, is probably not very well suited to take advantage of the new opportunity. But a juvenile, a larva, might, if its development could be arrested somehow, be much better equipped to found a new line of development. Adaptation fits an animal for a particular way of life. Changes are hard to accomplish successfully. But mutations that affect the early development of an animal are likely to set it off down a new path and enable it to exploit the opportunities that exist, and neoteny seems to have been responsible for some of the major evolutionary jumps, even though it is almost always an unsuccessful mutation.

Almost always, an infant that fails to develop properly into an adult will die a failure, without ever reproducing. But once in a million times, or even less often, the neotenous infant will fall on its feet, finding an ecological niche where it can survive, reproduce, and set life off down a new path of development. A process, like neoteny, that can keep throwing up cheap modifications of an existing line to test the water for new models must be efficient in evolutionary terms, and all that neoteny requires is a minor modification to the genes that control development. Simply by changing the tiny part of the DNA blueprint that specifies to the developing animal when to start and stop growing, a new line may be founded. There are, in all likelihood, only a very few genes that control the unfolding of development, so mutations to those genes will be quite rare. But when they do occur they are likely to have a profound outcome, affecting, as they would, all aspects of growth, maturation and development. All that external change balanced on a tiny internal change to the genetic code. The DNA of an axolotl is almost exactly the same as that of a normal salamander; the only significant difference is that the axolotl doesn't use its entire blueprint. And the DNA of tadpole and frog is absolutely identical. We have already seen that the differences between the DNAs of man, chimp and gorilla are minuscule, so small that many people are baffled by how they could produce the obvious and considerable differences in the external appearance of ourselves and the other apes. Neoteny resolves the problem. The one per cent difference could easily reside in the genes that control the rate of development, making human beings a form of infant ape that has learned to reproduce without reaching physical maturity.

At this point we should stress again that as well as being directionless evolution is absolutely not a case of one modern-day form evolving into another modern-day form. Palaeontologists are happy to talk about the similarities between a chordate and a sea-urchin larva because they know, from the fossil records, that ancestral sea-urchins, all but indistinguishable from modern species, were plentiful 430 million years ago at the time the chordates first appeared, but they do not believe that the sea-urchin larva that reproduced neotenously to give rise to the chordates was necessarily exactly like a modern larva. Similarly, we are not concerned with changing a man into a chimpanzee, or a chimp into a man, but with how both variations on the ape theme could have evolved from a common ancestor in the recent past just 4 million years ago. Nevertheless, it is easiest to grasp the problems, and highlight the evolutionary processes involved, if we compare modern man and modern chimp. In any case, we don't actually have any fossil evidence to tell us what their joint common ancestor was like, so we have to make do with the living forms.

You can see immediately, from superficial appearances, that man is the odd one out among the apes. He is naked, they are hairy; he has a nose, they have muzzles; he has a round, domed skull, they have flat ones; his big toe cannot be used like a thumb to grip, they have 'thumbs' on hands and feet. This is just part of a list of differences between man and the other apes that evolutionary theory must account for. The conventional contemporary wisdom, as we described in Chapter Three, is that man evolved from an ape-like ancestor that was not very much like a modern chimp. It was ape-like, but there are no fossils to tell us whether it had many of the anatomical hallmarks of the modern ape; probably it did not. The contemporary wisdom also has it that this ape-like creature became man-like as a result of neoteny.

It was a Dutch anatomist, Louis Bolk, who offered this new and daring theory to the scientific world, back in 1926. Bolk said outright that the human species is an extremely neotenous form of ape: 'man, in his bodily development, is a primate foetus that has become sexually mature.'[1] Bolk got his ideas from thorough and painstaking studies of the anatomy of man, other primates and other mammals; no clue to the very close genetic brotherhood of man and the African apes existed at the time but it seemed obvious to Bolk that although we do not have many superficial features in common with adult apes we do resemble ape infants. Bolk's list of similarities is long and impressive and a few examples from its score or more points is usually enough to be convincing.

We have already mentioned the large brain, housed within a domed skull. Other adult animals have brains that are, relative to their body size, smaller by far than man's. But the young, and especially foetuses, of all animals have both the domed head shape and the large brain (relative to body size) that we tend to regard as a distinctly human feature. It *is* a human feature, but it is also a feature of young animals of all kinds; how better to explain this than as evidence of neoteny in humans? At the start of an animal's development its brain grows especially rapidly, and the embryonic head is vaulted to accommodate it. Later on, in all species except man, the brain growth slows down as the body catches up, and the skull ends up flatter. Our brains keep growing for much longer, and stay large in proportion to body weight for life. When a rhesus monkey is born, its brain is already two-thirds of the size it will reach when adult; a new-born chimp's brain is two-fifths adult size; a human baby at birth has a brain only one quarter the size it will ultimately reach. The same pattern continues after birth: the brains of chimp and gorilla reach 70 per cent of their final size by a year after birth, but a human infant does not pass this milestone until after its second birthday.

The reasons why this evolutionary development is an advantage for our species are straightforward, provided we accept that bigger

(156)

brains are a good thing. The human female walks upright, and she has to be able to deliver her child. The hips must be fairly narrow for efficient walking, but the pelvis must be wide to allow for easy childbirth, so that in the female the two needs of locomotion and motherhood put conflicting demands on the design of the hips and pelvis. Calculations based on the fossil evidence tell us that this may be one area in which modern woman is less adept than her forebears, who, without the need to give birth to a big-headed baby, had a pelvis better suited to walking. Indeed, that is why men and women walk differently, because the modern female pelvis is more of a compromise, slightly less efficient than a man's for walking but able to do what no man could, allow a new human being out into the world. The pattern of growth of the human baby's brain is another compromise. The head at birth has to be small enough for the baby to emerge into the adult world without destroying its mother, yet the adult brain has to be large to take advantage of the evolutionary opportunities offered up by true intelligence. So a human baby is born premature, with a brain that is far from fully grown and a skull that is not yet completely covered in bone. Most animals are born with a completely ossified skull, a very important safety feature as they stumble about testing their legs, but the human baby has a large soft spot in its skull, and it takes years for the bones of the skull to close completly and become hard. This allows the brain to continue growing, with all that such growth implies for human inquisitiveness, versatility and intelligence, but it also means that babies and toddlers need much more protection while they are growing. Humans have to live in social groups, where adults care for and protect the children during their long period of development, for although they have left the protection of the womb they are still very vulnerable.

The complex web of evolutionary interactions is impossible to unravel completely at this level, and indeed it may be unwise to attempt to do so, because the entire nexus of humanity evolved as a coherent whole. Asking whether social groups came before the human pattern of brain growth is as futile as asking whether the chicken preceded the egg. We know that walking upright came first of all, because Lucy has a 'modern' pelvis and an 'ancient' skull, but who can say whether the pattern of brain growth had not already begun to alter in favour of delayed development. Many selection pressures were operating on the joint ancestor of ourselves and the apes to push them in the direction of neoteny. It is because neoteny answered so many evolutionary questions in one package that it was so successful, and it is the overall package that is crucially important in producing human beings and human culture.

Other features of the human skull and face are also juvenile, in the literal meaning of the term. Our face is flat, with no ridges above the eyes, no muzzle, small jaw and teeth. Human adults have a baby-face compared with other apes, and a baby chimp looks more like a human than like an adult chimp. Even the position of the human skull on the body is neotenous — the hole by which the spinal cord enters the skull is below the skull and points downwards in man, so that as we stand on our two feet our eyes are pointing forward. In other animals the corresponding hole is at the back of the head, where it has to be if the head is to point forward when the animal is standing on all fours. The advantage of our design, with eyes mounted at right angles to the spine, is obvious given our method of locomotion, and even more so because it allows us to carry our large brain around without having to use a great deal of muscle power to balance it. That is why the pattern evolved successfully, because it is advantageous. But *how* did it evolve? In other animals the position of the head in the embryo is initially the same as ours, bent over at the top of the spine, and it only shifts into the typical animal position later. The correct postioning of the large human head for walking on two feet is achieved by neoteny — the angle of man's head on his neck is the same as that of an embryo ape.

If you were to read a score of learned tomes on human evolution, you would find that they all agree that three features — large brain, small jaws, upright posture — are the most distinctive marks that set man apart from the other apes. All three owe part of their existence to neoteny, although as we have seen the upright posture was more the product of a history of brachiation than of neotenous development. Other features are less often mentioned as marks of man, but they are none the less highly important in moulding the social behaviour of the human species. The human vagina, for example, points forward, which is why we copulate most comfortably face to face. In other apes the vagina points forward in the embryo but shifts to the rear during the course of development. It would be fascinating, but not strictly relevant to our theme, to ponder the importance of this surely accidental and minor part of the neoteny package in establishing human social relationships. But the point is made, and

(158)

there is no need to labour it further here; in very many, though not all, respects a human adult is indeed a foetal chimpanzee.

All this evidence comes from a study of anatomy and growth, but it fits in beautifully with the fact that the genetic blueprints of man and the African apes are so very similar. Chimps are 99 per cent the same as humans, so whatever it is that makes us different anatomically resides in just one per cent of the DNA. The enormous superficial differences between ourselves and adult chimps and gorillas are misleading, just as a visitor from another planet might be misled by a tadpole and a frog. At first sight the two seem to be different species — one is a kind of water-breathing fish with a long tail while the other is an air-breathing animal with four legs and no tail. We can imagine some hypothetical space traveller who had never come across a life cycle like the frog's being astonished to discover that tadpole and frog share 100 per cent of their DNA.

Of course they do, they are different forms of the same individual. Yet the anatomical differences between tadpole and frog are even greater than the large differences between man and chimpanzee, who share 'only' 99 per cent of their DNA. Neoteny solves this puzzle. We are to the chimp or gorilla what the tadpole is to the frog — or the axolotl to the salamander. But this does not tell us *why* a man-like creature should have evolved from an ape-like ancestor. Some of the benefits we have hinted at — large brains and so on — but what environmental pressures triggered the shift?

There have been many theories offered to account for the emergence of the human species, and none of them is entirely satisfactory. Perhaps no truly satisfactory theory can ever emerge, because each has to be built on such flimsy evidence, the tantalisingly sparse fossil record. This record provides a glimpse of the animals and plants existing at a particular time and hints at the physical conditions, and in the right hands can be made to yield an enormous amount of information. But still our interpretation of the evidence is coloured by our own experiences, and depends on inner vision, imagination and speculation, hardly a sound basis for a scientific theory. Once we have built a theory — and this is as true of other sciences as of palaeo-anthropology — we are reluctant to change it. The human response to new evidence, if it cannot be slotted into the framework of our favourite card house, is all too often to ignore it. This tells us a great deal about our nature today, but very little about how we came to have that nature.

The true scientist, as portrayed in myth and legend, would always be ready, as T.H. Huxley used to say, to sit down before the facts as a little child, prepared to throw out even his best of hypotheses if the facts tell against it. In practice, most theorists cling too long and too tightly to their own brainchild and defend it against all criticism, so it is important to recognise that *no* theory of human origins is adequate. Our own ideas certainly are not offered as gospel truth, the result of some blinding revelation or new insight which henceforth must be treated as the only true word. We offer them simply as a possibility, a rather better picture, we think, than anything else yet developed simply because we incorporate a great deal of the latest evidence, the key features of which are neoteny and the very close similarities between human and ape genetic material. We expect our picture to be at least modified and perhaps superseded entirely as even more information becomes available, but we feel strongly that it is our duty to put forward this new picture as an antidote to some of the older ideas which are now thoroughly refuted by science but which refuse to relinquish the remarkable grip they have on the popular imagination. Many ordinary people, reading the works of a single populariser, believe that the riddle of human origins has been solved. In no case is this true, and all the ideas in print today — including our own — are more or less naked speculation.

Perhaps the most widespread and popular of these speculations presented as truth is the Mighty Hunter hypothesis. Adhered to once by hordes of eminent palaeontologists and other academics, popularised by excellent communicators such as Robert Ardrey, this view of human origins portrays a tree-living primate coming down on the plains as the great Pliocene drought got under way about 3 million years ago.[2] These early men carved out a new life for themselves by the use of weapons, forced co-operation, and unmitigated aggression. The appeal of this story was enormous, for if aggression made us what we are then we need feel no guilts about our increasingly violent behaviour. 'The human being in the most fundamental aspects of his soul and body is nature's last, if temporary, word on the subject of the armed predator,' Robert Ardrey told his huge audience. 'Man is a predator whose natural instinct is to kill with a weapon.'[3] According to the Mighty Hunter's advocates, the value of an upright stance was primarily that it enabled men to run fast and capture prey, not that it enabled him to escape or provided a good look-out for danger. Hairlessness became essential to allow the speedy runner to keep cool. Even human sexuality is explained by adherents to the Mighty Hunter hypothesis: sex becomes a way of keeping the human social group together, the male's unusually large penis and the female's orgasm — both distinctive human traits — combining to reward the female, while frontal copulation made sex more personal and ensured a strong bond between the pair. And breasts are 'explained' as developing from flat teats into pseudo-buttocks to replace the familiar sexual signals that males 'knew' from their days of mounting from the rear.

As we will see, many of these explanations are no more than special pleadings for man, and the facts are against the Mighty Hunter, and his champions. Set against their breathtaking vision of our past the evidence that the foot is better adapted to walking rather than running seems pedestrian in more ways than one, while the explanation of human brain development as a response to the need for a former vegetarian to learn what to do with the meat it stumbled across seems positively mundane. Yet here, surely, is the weakest link in the argument. Human success, over the past few million years, can be clearly related to our unique ability to be versatile, making the best of whatever opportunities came along. The Mighty Hunter theorists, by contrast, would have us believe that with one bound a timid ape freed itself and learned to compete directly and on more than equal terms with creatures such as the big cats and hunting dogs, genuine predators who had been evolving for millions of years into the hunting niche and whose genes already knew all about co-operation and the strength of the pack. And the prey animals — antelopes, gazelle, and so on — had not been idle either. They too had been evolving over a long period of time, becoming specialists in escape from increasingly efficient and cunning carnivores. To put this into perspective, a running man moves at about the same pace as a startled chicken, while a real sprinter like a cheetah or gazelle can manage speeds of 60 kph without too much difficulty.

The situation is even more complex than it appears at first sight, because even those big cats and other carnivores are not such mighty hunters themselves. The prey animals are wary, and very good at running away, and most carnivores are nothing more than glorified scavengers. Time and again, as many studies in the wild have shown, they pick off old, or sick, or very young members of the herd, the stragglers. Even though the hunter could almost certainly overcome a fully fit adult prey animal, the chances of it being able to catch such an animal are slim indeed, so it seldom makes the attempt. All this makes sound evolutionary sense.

A hunter that goes for the strongest, fittest member of the herd may get a bigger meal for its efforts, but it faces greater risks too. If it can catch a fit animal, which is doubtful, it stands an increased risk of being injured itself in overcoming its victim. There is a real risk of damage which will make the hunter a little slower and less able to make the next kill, which in turn decreases the hunter's chances in the struggle to survive and reproduce. Consistently, over many generations, the evolutionarily successful predators – the ones who leave most offspring to carry their genes on down the generations – are the ones who turn away from the healthy animals and concentrate on the weaklings that pose no threat whatsoever. A scavenger is a predator who eats carcasses that have either been left by the killers or bodies that have dropped dead through exhaustion or sickness; the mighty hunters are no more than glorified scavengers because, by and large, they simply pull out of the herd those bodies that are still alive and walking but only need a nudge to become carcasses.

So the true role for man on the plains was not as Mighty Hunter but as Mighty Scavenger – and that is where intelligence, rather than speed or strength, can be a huge advantage. Any ape might stumble across a dead or dying antelope, perhaps the remains of a mighty hunter's meal; but only a person would take a sharpened stone to cut meat from the carcass, and carry the meat back to base camp to feed the family.

A few apes, pushed out of their specialised habitat in the trees, would make a small impression as Mighty Hunters in opposition to specialised carnivores, even if armed with a rock or two to throw, for we are talking about times long before the invention of the spear, the bow, and all the other weapons which today make man the mightiest hunter on Earth. It just wouldn't work, and even a moment's cool reflection shows that it wouldn't work. But few stopped to reflect.

The Mighty Hunter hypothesis is no longer in great favour with professional anthropologists, but in its day it had enormous appeal, and it still appeals to a great number of people, especially those who see the hunt (or its bowdlerised modern substitute, sport) as symbolic of the virile activities beloved of true men. Like all theories of human origins — and we don't excuse our own — it tells us more about the people who stick to it today than it does about our ancestors in the past. Rather than tinkering with the old theory and adjusting it to fit the conflicting evidence — and there is plenty of that — what happens if we throw it out all together and start again, trying a completely different approach to the puzzle of how our ancestors became human?

In 1960 Sir Alister Hardy, Fellow of the Royal Society and professor of zoology at Oxford University, startled the scientific world — and the popular press — with his claim that our line began its evolutionary journey to humanity at the sea-side. Sir Alister was talking to a meeting of the British Sub-Aqua Club in Brighton, and while much of what he said was ignored, the suggestion that man evolved in the ocean was taken up with alacrity, at least for a week or two, and Hardy decided to write an article for *New Scientist* magazine to clarify certain misconceptions that had crept into reports of his ideas.[4] His fresh look at human evolution accounted, as we will see, for many human characteristics by showing how they could easily be adaptive responses to an aquatic, or semi-aquatic, way of life, and although largely ignored by mainstream academic palaeo-anthropology, Hardy's theory provoked author Elaine Morgan to write her best-selling book *The Descent of Woman*, which appeared in 1972.[5] Morgan built on Hardy's theory, but from her own female point of view, with the result that *The Descent of Woman* provides almost a total antithesis to the Mighty Hunter hypothesis. Instead of man as the key element in human evolution we have woman; instead of the scorching plains as the site of the major changes we have the ocean; and instead of aggression and hunting as the driving forces we have domesticity and the comforts of home. Again, this story was massively popular, the more so given the burgeoning consciousness of young women of the time. Morgan provided an antidote to male-dominated origins, and she included a great deal of good old-fashioned sex to help her argument. Face to face copulation, Morgan said, was a legacy of our more recent evolutionary past that we hadn't fully caught up with; it was responsible for almost all of woman's woes. As a result of this curious posture, she said, the penis applied pressure to the wrong parts of the vagina, thereby robbing modern women of the orgasms that nature had intended. Of course the idea was a success, catering to an almost tangible intellectual need. But to anyone seriously interested in where we come from, a study of the two opposed views is a real eye-opener, because if two such different models of human evolution can be built using exactly the same available evidence it becomes abundantly clear just how little we really know about human origins.

Morgan's view, extending Hardy's original thesis to encompass just about every aspect of human anatomy and behaviour, surely goes too far in trying to explain everything through a solitary mechanism. She certainly starts on solid ground, with her contention that a tree-dwelling ape would not stand a ghost of a chance on the open plains in direct competition with the real mighty hunters like the lion, the hunting dog, and the hyena. And, following Hardy, she develops the idea that an ape that came down from the trees not to the plains but to the water provides a much better starting point for the road to humanity.

The basic premiss is that about 12 million years ago some apes were forced by the furnace heat of the Miocene to move to the sea-side, where they could keep cool and where the ancestral proto-woman could escape from predators by wading out into the shallows and waiting until the menacing beast got bored and went away. All else follows. Upright stance enables her to wade out further; hair is redundant, a positive hindrance to an aquatic mammal especially in the warm conditions of the Pliocene. Hair on the head remains, as protection against the sun and as a safety net for the infant to cling to. Fascinatingly, the residual hair on the body today, and even more so the hair on a new-born babe, is distributed in a pattern unlike all other primates' fur but like the streamline pattern of water flowing over the body; this is an impressive plus for Morgan. Without hair, swimming would be easier but some other form of insulation would be needed, which explains why people, especially women, have evolved a blubbery layer of fat just under the skin, like the porpoise (a mammal that definitely returned to the sea) but unlike every other modern ape. The same blubber would provide a cushioning pair of buttocks when sitting on rocks rather than grassy banks and trees, with the female's vagina receiving further protection as a consequence of its more forward position and the development of a hymen; face to face copulation and the male's

(164)

large penis are an inevitable result. Woman's breasts evolve not to titillate her mate but to enable her child to reach the nipple and suckle without the aid of body hair to cling to; a baby nestling in the crook of a mother's arm is out of reach of the food supply, unless the food supply is lowered on a breast.

Many of these features, as we have seen, can be easily explained in terms of *how* they evolved by neoteny. The argument is about the *why*. And it doesn't matter which side of the argument — if either — you may support as far as the neotenous nature of human development is concerned. Neoteny is established beyond reasonable doubt, and perhaps one of the things that the success of both the Mighty Hunter hypothesis and the Aquatic Woman model tells us is just how useful an adaptation neoteny is. It could have helped in a hunting life on the plains; it could have helped in a semi-aquatic life by the waterside. Most of all, by freeing us from the trap of adult specialisation, it helps us to be versatile, the key to our own view of human evolution. Neither Mighty Hunter nor Aquatic Woman is entirely satisfactory as an explanation of everything human, and we can perhaps highlight the difficulties involved in constructing any picture of human origins by confessing that the two of us do not agree on how much can be saved from these two widely — even wildly — differing perspectives. We both agree that the Mighty Hunter can be discarded utterly. For the reasons we have outlined, and which Morgan elaborates on, it makes no sense to see human evolution as the outcome of a dramatic new specialisation by a forest ape in competition with much more efficient specialised hunters. The Aquatic Woman has a lot more to offer, but we cannot agree how much more.

Her first great advantage is that she is part of a much more generalised model of evolution. Tool use, for example, develops gradually in this picture as a means first of cracking shells open to get food, and fingers become more sensitive as they are used to probe for food in the mud. This makes a lot more sense, as a response to steady evolutionary pressure, than throwing rocks to bring down game, literally a hit or miss affair with little opportunity for learning if you get it wrong. One of us thinks that the best one can say of Morgan's version of Hardy's hypothesis is that it deserves to be taken seriously only as a sensible example of model-building, unlike the Mighty Hunter hypothesis which is useful only as an example of how *not* to construct a scientific theory. But he doesn't think that there is more than a grain of truth to Morgan's version of the story of human evolution; it took place by the water's edge, not to escape from predators but simply because this was a good place to make a camp. The other author, however, finds the aquatic evidence too strong to dismiss entirely. The streamlining of body hairs we have already mentioned. In addition, the human nose — a very peculiar feature for a land animal, as a visit to any zoo will testify — is a rather good device for keeping water out of the sinuses and lungs while swimming. If nothing else, this idea makes more sense than Sir Fred Hoyle's bizarre notion that the human nose evolved to prevent bugs from space falling into our nostrils.[6] More abstractly, the human frown — a feature that puzzled Charles Darwin because no other primate seems to use the muscles we use to frown — could be an evolutionary response to the need to shut out some of the glare of sunlight reflected off the water. And more concretely, why is it that all newborn human babies can swim? Furthermore, it has been claimed that man is unique among the primates in possessing a complex of adaptations that physiologists call the diving reflex.

Mammals that have re-adapted to a life in the oceans, for example the whales and seals, have evolved a special mechanism that allows them to make long dives in safety. When they go deep underwater, a reflex mechanism slows down some bodily processes and reduces the supply of blood to parts of the body that can do without for a while. At the same time, other reflexes ensure that there is a good supply of oxygen-carrying blood to the brain, which cannot do without. Coupled with the extraordinary affinity of their haemoglobins for oxygen, which means that they can use the last available drop, these diving reflexes enable sea-living mammals to exploit the oceans fully; of course they have to come up to the surface to breathe, but to all intents and purposes the cetaceans — whales, dolphins and porpoises — are as effective in water as fish, ensuring that their brains remain supplied with oxygen even on long, deep dives. A human diver who has been properly trained not to panic at the first feeling of breathlessness can do much the same sort of thing on a more modest scale; after the breathless stage the body seems to enter a calmer state apparently adapted to underwater swimming, with blood shunted preferentially to the brain. The diving reflex is now well established as a feature of human biology and it has revolutionised the way we teach first aid. Until a few years ago, would-be rescuers were taught that if a drowned person had been underwater for more than a couple of minutes it would be kindest not to attempt a resuscitation, because even if breathing could be re-started the victim would have suffered permanent brain damage as a result of oxygen starvation. Today, the received wisdom is that one should always try to resuscitate apparent victims of drowning, and that, especially in cold water, the fact of the diving reflex provides a slender but real chance that even after several minutes of complete immersion the victim can be 'brought back to life' unharmed. Why, if there is not at least some truth to Elaine Morgan's version of human origins, would we have a superbly adapted set of physiological mechanisms like the diving reflex? On the other hand, while it may occasionally save his life, man's diving reflex isn't nearly as well developed as, say, a duck's. Furthermore, all reptiles, birds, and mammals that have been tested also show at least a partial diving reflex; we all did have an aquatic ancestor, but it was long before the Pliocene. The diving reflex is most easily interpreted as a mechanism to protect the young animal during its struggles to emerge from the watery egg or womb.

Hardy's list of human attributes, as extended by Morgan, is long and persuasive — longer and more persuasive that the equivalent list of the Mighty Hunter's advocates. Man does show a great many specialisations that otherwise only appear (to a greater extent than in man, of course) in marine mammals. And the theory accounts rather neatly for the gap in the fossil record before the appearance of hominids. There are no fossils, says Morgan, because our ancestors lived by the shore where bones were quickly destroyed or washed away, and there are no stone tools because it would be considerably easier to find a needle in a haystack than a pebble tool on a pebble beach. Unfortunately, though, the aquatic theory cannot explain everything about human evolution, and there are some glaring faults in it.

Hardy and Morgan, writing 10 or more years ago, got their dates wrong, of course, but so did everyone else in the 1960s. As more effort has been put into fossil-hunting, more — and earlier — remains have been found, so that this inaccuracy cannot be reason to discard the Aquatic Ape hypothesis. But it does narrow down the fossil gap that the aquatic model was designed to fill, leaving less time for sea-side evolution to work its changes. If the key event in shaping our ancestors was the great Miocene drought, 8 or more million years ago, then according to the molecular evidence the ancestor it shaped was also the ancestor of the chimp and gorilla — and Lucy. To explain features believed to be uniquely human as due to an aquatic phase of evolution, we have to move the aquatic phase forward to 3 million years ago, and it could scarcely have lasted a million years. There are, relatively speaking, plenty of recognisably hominid ancestors well inland during this time. More than anything else, this change of timescale weighs against the Aquatic Ape hypothesis. In any case, the great Miocene drought, it seems from recent climatological studies, wasn't all that much of a scorcher; the forests shrank, it is true, but the severity of the change probably wasn't as great as Morgan made out. More to the point, the Wading Ape model is as dubious as the Mighty Hunter hypothesis in its basic premiss. Morgan states baldly that a tree-ape couldn't survive on the plains but could survive in the shallows, where the carnivores could not catch them, and indeed she paints a vivid picture of the terrifying life on the ground for an ape. But is she right? Many carnivores, including the big cats, are able swimmers and would not hesitate to dive in in hot pursuit of a tasty meal. And she seems never to have heard of the legendary patience of the social carnivores — hunting dogs and hyenas especially — who are content to wait almost indefinitely if they know they have cornered their quarry, secure in the knowledge that the meal will be ready for them sooner or later. The greatest puzzle of all, though, is why creatures as superbly adapted to the aquatic life as Morgan would have us believe should ever abandon their new home. The theory does at least try to explain why tree-apes ventured into the water, but all it says about the U-turn that brought our ancestors back on land is that once the rains returned and the land became green once more the Aquatic Apes would have been tempted up rivers and back into the hinterland of the continent. But the greatest evolutionary pressures come from harsh circumstances that force species to adapt or die; merely opening up a garden of Eden suitable for land apes would not cause water apes to leave the shore, especially if there were already other animals adapted to the plains, as there surely were. The better the water-apes' adaptations to life by the sea-side, the less likely they would be to leave it; if the aquatic adaptations were not a great success then the theory falls down in any case.

From our own immediate viewpoint, of course, the main deficiency in the concept of the Aquatic Ape is that the events described so imaginatively by Morgan must have taken place more than 4 million years ago, that is before the split between man, chimp and gorilla. Ten years ago, Morgan could happily talk about the apes of that time as uniquely human ancestors, accepting the almost unquestioned view that the human line had been distinct from the ape line for more than 20 million years (although by 1972 the first hints of the recent time of the man-ape split had been in print for over 5 years). Today, we know better. If man has aquatic features left over from an ancestor alive 4 or more million years ago, then chimp and gorilla should share those aquatic features, although of course we would not expect them to be identical. Nobody has made an in-depth study of the aquatic skills of chimp or gorilla — perhaps someone should — but, as far as we know, the evidence is that man is the more aquatic of the three. This hints, perhaps, that swimming and camp life by the sea or lake shore have played a significant part in the lifestyle of our ancestors since the time of Lucy. And that means that a waterside lifestyle can only have played a part in our evolution after the crucial steps that set us on the road to humanity.

What we finally come down to is the question whether life on the savannah would really have been so very difficult for our ancestors. We think not, especially if they avoided competing directly with the real mighty hunters. We have only to glance at the plains today to find many primates, including man himself, doing very nicely, thank you, and we are entitled to ask how baboons and vervet monkeys could manage today in circumstances that their, and our, primate ancestors allegedly found so inhospitable. The sea-side model of human evolution seems not to be needed in its complete form because although it is superficially plausible and apparently accounts for many of man's adaptations, the original premiss — that our ancestors could not survive on the plains — does not, in fact, hold water. There is, however, still scope for enthusiastic supporters of the idea to argue that our ancestors may at some time have been involved in a more watery way of life than that of modern man, although they cannot be as firmly committed to the all-explanatory power of the hypothesis as Morgan, and they cannot push the watery phase back so far into the mists of time.

Enthusiastic supporters of the Mighty Hunter hypothesis, on the other hand, have even less solid ground on which to stand, and fewer fragments of the hypothesis can be rescued to be incorporated into a more generalised picture of human origins. If hair was lost to help the hunters cool down after the strenuous chase, why do men, the hunters, have more hair on their bodies than women, who allegedly stayed on the sidelines watching? Could tools have begun as weapons? If you throw a rock and miss, quite likely for a beginner, you either go hungry or get eaten, depending on your target, with very little scope for trial and error learning. We can learn from modern primates too. Baboons do not compete with lions, but follow their own lifestyle; if our ancestors had been successful competitors with the big cats, natural selection would have taken them down a path very different from that leading to modern man. Some of us like to think that we are fierce aggressive animals, but in truth we are pretty pathetic as fierce animals, stripped of our technology. Over millions of years the evolutionary needs of a hunting life could at the very least have equipped us with the ability to run really fast, or a decent set of teeth, and would surely have retained a far better sense of smell and taste. Instead, we went down a path marked by versatility, adapting our environment to suit our needs, with the result that there are modern men who are indeed fearsome hunters, but there are also modern men who couldn't catch an animal even if their life depended on it. We can do anything we set our minds to, which is not the mark of a Mighty Hunter but of a Mighty Thinker.

Interpreting fossils is a guessing game, and for any question you like to ask it is usually possible to come up with more than one answer. Even then, none of the answers may be correct, and the fossils may not be telling the truth. Accepted dates change, and new evidence comes to light. We now know, for example, that although the Miocene-Pliocene boundary, 5 million years ago, was a time of drought and heat in Africa, compared with conditions 20 or 30 million years ago, the African savannah was not very different then from the savannah now — there was just more of it. Rivers still ran, and trees grew along the rivers. The climate was probably more variable then than it had been before, more likely to produce an extreme local catastrophe and force animals to move, adapt, or be wiped out, but these are the very conditions that would encourage evolutionary diversity and change. This picture too may be improved as new evidence comes to light, but at present the best evidence we have is that the start of the Pliocene was neither so harsh as to wipe out everything not already living in semi–desert, nor so swift in arriving that most species died out without having produced descendants that were better adapted to the new conditions.

Theories are not sacrosanct, they must be changed as new evidence comes in. The way we interpret history and pre-history must be malleable, altered when made necessary by new evidence or new ideas, and that is why we think Hardy's hypothesis of the sea-side ape was a good one. We don't think it was correct, but it gave new insights and spurred people on to produce a better version than the Mighty Hunter hypothesis, the only alternative at the time. Even though incorrect, Hardy's idea is not a bad hypothesis, for it showed how truly bad, by comparison, were the alternatives.

The major fault of both Mighty Hunter and Aquatic Ape is that they tried to do too much too soon. Each looked for a single event that set a tree-ape on the road to humanity; in one case life in the ocean made the difference, and in the other hunting big game was the key. Of the two, life in the water can explain more, but even that is precious little, and in any case there is no need to make a choice between the two because there is no need to invoke a single pre-historic event that precipitated all the events that lead to modern man. Different evolutionary pressures resulted in different adaptations, and the end result is modern man, but the key feature of man remains his versatility. He is jack of all trades, going in for a little bit of hunting, a little swimming, a lot of farming, and so on. No one step in our recent evolutionary development dominated, and if it had we would not be here now — we would be either pseudo-wolves or pseudo-dolphins, social but specialised. Hunting may have played a part in the general pattern of human evolution; life by the shore probably contributed at some stage; but neither was the single dominant factor in shaping modern man or woman. It is interesting, too, that the very latest reconstructions of the life of early man invariably include a home base by the water's edge and a bit of hunting by the men; the best of both worlds.

Let us be frank. Nobody knows what happened during the Pliocene to make us human, the chimps chimps, and the gorillas gorillas. We know that, for us, the first step was to walk upright; big brains came later. But we don't know why it was a good thing for our ancestors to stand up, or exactly how they benefitted from big brains. We can make some good guesses, of course, but this still represents a huge gap in our understanding, for although many features of the human species are the product of neoteny, upright walking certainly is not. The foetus's legs are small and underdeveloped compared to its arms, reflecting perhaps the brachiating way of life of our ancestors. In any case, neoteny explains how, not why, we changed. Our own version of the story of human development is an amalgam of several stories, based on the unambiguous evidence of molecular anthropology and using the available fossil evidence as interpreted by the experts, and it is certainly less dramatic and exciting than the black-and-white debate between Aquatic Apes and Mighty Hunters. It is, however, a plausible story, and even if it is overtaken by new discoveries it provides an instructive way of thinking about our ancestors here and now.

For the record, we think that man evolved on the savannah, and that he also spent a great deal of his time camped in family groups by the water's edge. But, in common with Richard Leakey and Don Johanson, we do not see our ancestral way of life on the plains as that of a mighty hunter, but rather a wise shopper. Glynn Isaac, a brilliant anthropologist at Berkeley in California, conjures up just this image with his claim that the most important advance of all time, the single invention that created man, was the basket. What he means is that the basket — something to carry things in — enabled early proto-man to develop the society that eventually led to modern man.

It is easiest to see why the ability to carry things is so crucial if we consider some of our relatives. Baboons and vervet monkeys, though a great deal further from us than chimps and gorillas, live on savannah not much different from the savannah that our ancestors inhabited 4 million years ago. A large part of their day is spent in the vital business of getting the food needed to stay alive, time spent either in actual feeding or in moving from one feeding site to another. Each monkey fends more or less for itself, and although they have a rich and entertaining social life it is nowhere near as complex as even the simplest human society. The reason is that they just do not have time for this complexity. Even the least technological of human beings — the bushmen of the Kalahari desert or the aboriginal Australians — build a complex society, and they do so largely around food. They make camps that they occupy for several weeks at a time, and they return each day from foraging expeditions bearing food. It isn't a hard life by our standards, just a couple of hours a day gathering roots, berries, grubs, nuts, and whatever else they come across. Mostly this gathering, which is the major source of protein and energy, is woman's work, something she can do quite easily with a small child in tow, provided that she has some receptacle in which to carry what she has found. Two handfuls

— which assumes a child that is either independent or can cling, a poor assumption — are not worth the journey to collect them. But a basketful is quickly gathered and carried back to camp, making this way of life relatively profitable and providing the foundation of society. The men do hunt, it's true, but this is almost a joke compared to the hunting lifestyle of a true carnivorous predator. They don't hunt often or catch much, and the lion's share of the food is provided by the women from their baskets when they return to camp. ('Lion's share' is entirely appropriate; in a pride of lions the females are usually responsible for the kills and the males drift along once all the hard work has been done and eat their fill. The lion's impressive mane is excellent for showing other males what a fine creature he is, and may play a part in his sexual success, but is hardly ideal camouflage for stalking wary wildebeest.)

This pattern — women gathering and men hunting — is found so clearly among so-called primitive people today that it makes sense to see it as akin to the ancestral way of life of plains-dwelling proto-man. Women today are often quite rightly incensed by the neglect shown to their sex in older stories of human evolution, and they have a very strong case for arguing that they played the more important role in the crucial developments that made us human, even though those developments did not take place exclusively at the sea-side. And for those who like to conjure castles in the air from the slenderest of building materials, it would be possible to argue that man's rather feeble efforts at hunting developed not so much because the poor woman and her children would starve without red meat provided by their big strong protector but simply to get him out of the way while she got on with the important business of raising children and gathering fruits, nuts, berries, and so on. It gave the men something to do, brought in a little meat as a bonus, and kept them out from underfoot. It could be for the same reason — because they had time on their hands — that men became the custodians of tradition, the wise old men of the tribe and the tellers of tales. The ironic joke, as far as women are concerned, is that men remained the tellers of tales right into the final fifth of the 20th century, and one of their tallest stories was the one about how man the Mighty Hunter was such a success on the plains of Africa millions of years ago. The pattern continues today, with man the breadwinner staggering home after a hard day at the office, where he signed a few letters and chatted with his cronies, only to be waited on hand and foot by

(173)

woman the home-maker, who has merely had to look after house and children, find and prepare the food, and pander to her protector. It is an ancient pattern. Our Mighty Hunter ancestors probably did as much mighty hunting as the average filing clerk does today.

Nor is this braggard behaviour entirely purged from the halls of academe. One respected anthropologist — Owen Lovejoy of Kent State University in Ohio — recently suggested that males were the ones who learned the clever trick of walking upright, a proposal taken up enthusiastically by Don Johanson.[7] The argument, crudely, is that a male who could walk upright would be able to carry food back to his mate and her children, and that such a male would raise more children than a male who did not walk upright and did not provision his mate. The female sat quietly waiting for food that the male brought home, and the male grew larger than the female to protect himself from the dangers he might encounter on his foraging trips. So while acknowledging the bankruptcy of the Mighty Hunter hypothesis, this notion seizes the primary role for the emergence of mankind and gives it back to men. There are so many conceptual faults with this model that it is hard to know where to begin criticising it, but someone who has taken on the task is Becky Cann, a student with Sarich and Wilson. She points out that one factor that weighs against males is know in the trade as paternity uncertainty. A man can never be certain that the child his mate produces is his own; cuckoldry is always a possiblity. A woman suffers no such doubts. For this fundamental reason natural selection will always be less efficient when working through males than through females; in the worst case the super-upright wonderful provisioner would be helping his wife to raise the offspring of some shambling philanderer and doing nothing to spread his own wonderful genes. Not a very salutary prospect. Cann points out that the most important social bond between humans is not that between mates but between mother and child, and stresses that food-sharing, while indubitably of very great importance for human development, was probably first shown between mother and children, as indeed it is in many other primates.

The mother who can carry her infant and gather food, as a bipedal mother would be able to do, would be better off and would indeed be more successful. Her sharing and caring would spread so that her older children would help to take care of their younger sibs, and thus the mother and her offspring would form the nucleus of the social group. Of course a mother might still want to be protected from the unwelcome advances of a male intent on furthering his reproductive success; what better way to do so than to form a long-term alliance with one male? There are many ramifications to all this, perhaps the most important of which is that it allows for flexibility in the way that human societies organise their relationships, something that Lovejoy's monogamous male provider theory does not. His theory is a direct descendant of the chauvinistic anthropological theories of old, a point brought forcefully home by Cann: 'In their eagerness to impose industrialized western cultural standards on the fossil record,' she says, 'some paleo-anthropologists . . . seem to have forgotten their own best advice, which was to remember that the present may not be the key to the past.'[8]

To return to our story, proto-people, out on the savannah, surely had brushes with predators, and some of them must have come off the worst and been eaten. But they learned caution and found secret places where they would be safe at night. The males and females might have gone out on foraging trips together, as modern baboons do, and they would have learned to eat meat when they found it, scavenging someone else's leftovers, perhaps, or taking advantage of an animal — young, old or sick — that couldn't escape. Tools — including the quintessential shopping basket — would have developed not as a single flash of inspiration (like the bolt of outside help in Stanley Kubrick's film *2001: A Space Odyssey*) but as slow improvements. A stick to grub up roots and bulbs; a rock to smash bones and release the tasty nutritious marrow; a splintered stone to cut up meat and make it easier to carry. With a central base-camp to return to each day, even a peripatetic camp, long-term bonds could develop between people other than mother and children. Individuals could begin to rely on one another, sharing skills, information and food. Mothers with very small infants, the old, the young and the sick could stay in base camp, safe from the attentions of other predators and tended by the fit and healthy. Gradually from these shared obligations and commitments the unique complexity of human society sprouted; the facts are slender, but it must have been something like that. The crucial point is that all the changes are interrelated. Food-sharing favours society, and society depends on sharing food. Walking upright helps to gather food, and gathering food fosters the ability to make primitive tools. Better tools are made by those with better brains, which also promote more complex relationships with other like-minded people, and so it goes on. The skein is a tangled one, but it acts as an amplifier, magnifying any small tendency towards the human way of life.

What we do know for sure is that primates were walking perfectly upright just less than 4 million years ago, long before the end of the Pliocene drought, and they were doing so well inland, not at the sea-shore. We also know that these particular primates had skulls that were very primitive, with brains that had not yet begun to balloon. We think they were omnivorous, eating whatever they could find. Probably the best modern model is the chimpanzee, which survives on fruit, nuts, seeds and roots, insects, and occasional meat. But we have no idea why they found it evolutionarily worthwhile to undergo the major bodily upheavals involved in standing upright, upheavals that were made easier by a previous life as a brachiator, but that were nevertheless considerable. It surely was not to carry weapons and chase prey; we are far too bad at that sort of thing today for it to have played a major role in moulding us. It might well have been an adaptation that enabled us to carry helpless infants and food, while at the same time providing a better look-out for the real mighty hunters of the African savannah.

One further piece of real evidence shows the importance of the shopping basket in providing an impetus to the evolution of modern man, and nowhere is the standard terminology 'man' to include both sexes less appropriate. The female human's pelvis and hips are structured as a compromise between upright posture and bearing young. As a result her arms would brush against her hips all the time if they hung straight down at her sides. But the elbows of women are constructed slightly differently from those of men, giving a slight outward kink which leaves the hands swinging free when the arms are relaxed by the side of her body. What possible adaptive significance, what evolutionary advantage, could be conferred by this subtle modification of the human elbow? Dare we suggest that the woman's arm is perfectly shaped for the job of carrying a heavy bag of shopping (or roots and fruits, it doesn't matter) without getting in the way of her hip? This speculation is emphatically not intended to demean the role of women in society. Far from it. For, as we've said, it was almost certainly the combination of females and their baskets that made us human.

Once again, though, we come back ultimately to neoteny as the key factor. No single speciality is seen as the driving force behind our evolution, but rather our general adaptiveness and versatility, the hallmarks of their infant ape. Babies we are for sure; water-babies we probably are not, but some of our ancestors surely swam in lake or sea, an activity all too often ignored by traditional palaeo-anthropology. By contrast, the conventional view surely places too much emphasis on the hunt, although again some hunting undoubtedly did take place.

Does the picture we have painted here help to resolve the monkey puzzle, of why we share so much of our DNA with the chimp and gorilla? Certainly neoteny accounts for the mechanics of the process whereby a minuscule change of just one per cent of the genetic blueprint can produce creatures as different in outward appearance as ourselves and the African apes. The final shape of an animal is dictated largely by the rate at which different parts develop and grow, and it is only the few genes that control the unfolding of development that would need to change in order to produce an upright ape from a knuckle-walker. Neoteny explains the whole package of anatomical distinctions that characterise man, with the exception of strong legs, and the changes to the genes to permit neoteny would not have taken millions of years to accumulate. It is entirely possible, indeed probable, that the quarter of a million years between the molecular clock's timing of the ape-man split 4 million years ago and the fossilised remains of Lucy 3¾ million year ago was more than long enough for these changes to have taken place.

Whether the palaeontologists like it or not, their cherished fossil timescale for human evolution must be wrong. *Ramapithecus* was not an ancestor of man, and the ape line certainly did not split from the man line 20 million years ago. With neoteny as the mechanism, the split could very easily have occurred 4 million years ago, as the molecules say it did. One of the frustrations engendered by the scarcity of fossil evidence, however, is that we have no fossil evidence at all of species that we could regard as the direct ancestors of the modern African apes, except for a couple of teeth so old that if they are ancestral to the apes they must also be ancestral to us. In circumstances such as these, any theorist is simply shooting in the dark. And, in such circumstances, we cannot resist a shot of our own.

The best scientific tradition involves the setting up of theoretical ideas as targets to be shot at and perhaps knocked down. This is often called the 'what if . . .' approach, and, somewhat facetiously, we can imagine Newton musing, 'What if the same force that pulls an apple off the tree also hold the planets in orbit around the Sun?' Most 'what if' hypotheses are eventually disproved — even the respectable ones — but the very act of disproving them adds to our store of knowledge and enhances our understanding of life and the Universe we inhabit. Quite recently, for example, cosmologists were faced with two rival hypotheses. One said, 'What if the Universe started out in a Big Bang, a single explosion at a definite instant thousands of millions of years ago?' The other said, in effect, 'What if the Universe has always been the way we see it now, always expanding but with new matter continually being created to fill the gaps left by expansion?' Both ideas fitted with the key observation, that the Universe is expanding; more than 20 years of conflict between the models encouraged an enormous number of new observations, using radio telescopes, optical telescopes, and a whole variety of other instruments, and the result of this endeavour is that today the steady state theory has been discarded, to the satisfaction of all but a few diehards. The understanding we now have of the beginning of the Universe is truly astonishing.[9] The point is that the new evidence that so strongly indicates that there was a Big Bang at the beginning was collected as much as a result of the existence of the steady state theory as because cosmologists had a Big Bang theory to examine. The conflict between the two models acted as a spur to intensify the observational studies that alone could provide a resolution to the conflict.

Our modest proposal is in this tradition. We do not claim to have any ultimate truth. When, and if, conclusive evidence turns up we will be glad to see it and will abandon this hypothesis happily. Until then, what if the man-ape split went something like this?

To keep the traditionalists with their fossil timescale reasonably happy, let us suppose that a primate clearly on the road to being human was present somewhere in Asia between 10 and 14 million years ago. This would be a descendant of *Dryopithecus* but not *Ramapithecus*, which is almost certainly a dead end. During that embarrassing gap in the fossil record, evolution proceeded, producing, by 4 million years ago, the upright proto-man and proto-woman with her shopping basket, successful inhabitants of the dry plains. Then, indeed, the climate did change for the better, resulting in a spreading of the forests from their beleaguered positions, more rain, more fruit to eat, and so on. What would happen to those proto-humans who did indeed follow the path equivalent to the one signposted by Elaine Morgan in her efforts to explain why her Aquatic Apes left the water? The sybaritic life — sitting in, or under, trees and munching fruit and vegetables — would hardly be likely to stimulate development of the brain. And with an abundance of goodies to hand the shopping bag would be neither useful nor necessary. Without the need to share food and rely on one another, what use is a complex society? You probably see where we are going: chimp and gorilla, it is possible, could be the remnants of two lines that gave up the hard life of the plains for the comfort of the forest, and they did so after man had taken several steps on the road to humanity. Now if changes to only a few genes can, through neoteny, turn a hairy tree-dweller into a naked, upright proto-man, then only a few small changes to the same genes can equally likely turn a naked, upright proto-man back into a hairy tree-dweller. Neoteny, after all, is in some sense an artificial slowing down of development; simply by taking the brake off and letting nature take her course, full 'normal' development would result. And the full 'normal' development of a man — the neotenous ape — would be something very like a chimpanzee or gorilla.

What we are saying is really an extension of Louis Bolk's theory. Modern man is the neotenous child of a more ape-like father; what if modern apes have let go of man's hard-won youth and reverted to that original ape-like form? After formulating this entertaining idea, we discovered that it isn't so original after all. Aldous Huxley (grandson of T.H.), whose brother Julian must have kept him well informed about matters scientific, including Bolk and his theory, did it first. He speculated along the same lines and worked his speculation into a novel, *After Many a Summer*. The hero, an English scholar called Jeremy Pordage, works for a Californian millionaire, Jo Stoyte, who fears death above all. Stoyte employs the mephistopholean Dr Obispo as his personal physician to stave off the inevitable. The evil Obispo is a fraud, but Pordage discovers an old diary which hints that the Fifth Earl of Gonister has discovered the secret of eternal life in the gut contents of the carp, a fish renowned for its longevity. After many problems, Stoyte, Obispo and their mistress decamp to England, where they go in search of the Fifth Earl. They eventually get to his country seat, and make their way down to the cellars, where a horrific scene awaits them:

'On the edge of a low bed, at the centre of this world, a man was sitting, staring, as though fascinated, into the light. His legs, thickly covered with coarse reddish hair, were bare. The shirt, which was his only garment, was torn and filthy. Knotted diagonally across the powerful chest was a broad silk ribbon that had evidently once been blue. From a piece of string tied round his neck was suspended a little image of St George and the Dragon in gold and enamel. He sat hunched up, his head thrust forward and at the same time sunk between his shoulders. With one of his huge and strangely clumsy hands he was scratching a sore place that showed red between the hairs of his left calf. "A foetal ape that's had time to grow up," Dr Obispo managed at last to say . . . Without moving from where he was sitting, the Fifth Earl urinated on the floor.'[10]

The Fifth Earl had indeed discovered the elixir of eternal life, but not eternal youth. He had outgrown his neotenous body and become the ape his ancestors had been, millions of years before.

What Huxley used as a novelistic device, as much to point up the futility of immortality as a theory of man's evolution, we put forward as a serious suggestion that is not contradicted by any evidence of which we are aware. We suggest, then, that the split between man and the other African apes did indeed occur within the past 5 million years, but that it did not occur in the direction that conventional wisdom implies. The last ancestor shared by man and the apes was already almost human and well adapted to life on the plains, and while many members of the species stayed on the plains, honing their skills and eventually becoming fully human, some tribes took advantage of a period or periods of climatic amelioration to choose the soft option of life among the trees and free lunch all around. The two-edged sword of neoteny helped to make proto-man in the first place, but it also made it easy for proto-man to discard his neotenous development and return to the full adult form, like the axolotl given a tiny dose of iodine. (There is a little evidence against our wonderful hypothesis. The pelvises of chimp and gorilla do not show any sign of having once been the pelvis of a bipedal species; but then, perhaps the reversion to an ancient type has been perfect.)

This proposal may seem like a wild flight of fancy, the overvivid imaginings of two armchair anthropologists with no expereince of palaeo-anthropology in the field. But while we certainly would not want to defend our idea to the death we can, perhaps, put it into a respectable context by fitting it into the story that Don Johanson tells about his most startling discovery, the 3¾-million-year-old Lucy and other remains from Ethiopia. The key insight that Johanson describes, the blinding flash of light that allowed him to see how the Hadar fossils, including Lucy, could be slotted into place on the monkey puzzle tree was, he says, the realisation that the teeth of the fossils on the *Homo* line had changed very little over the period from 3 million to 2 million years ago. Previously, along with other palaeo-anthropologists, Johanson had been working on the assumption that large teeth were more primitive and that fossils with smaller teeth must be more evolved and hence more recent, the result of changes related to a changing diet. 'I now saw,' says Johanson, 'that what I had taken for a late human trait was actually a primitive one. A better word here would be "old", because primitive suggests something less good, less highly evolved, whereas in truth it may be perfectly good.'[11] Once he had realised that the teeth of fossils on the main *Homo* lineage at this time had changed very little — presumably because they were well adapted to the diet of our ancestors and had no need to change — it was apparent to Johanson that the changes in the fossil record showed only that the teeth of our cousin hominids, the australopithecines, had altered. 'They had gone in a direction of their own,' as he put it, 'to satisfy a life-style somewhat different from that being lived by early humans — a life-style that would become increasingly specialized and lead to the development of larger and larger teeth.'[12]

Lucy and her kindred become, on this picture of Johanson's, the ancestor common both to the *Homo* and *Australopithecus* lineages, the last of our non-*Homo* ancestors. This sets the date of the oldest 'human' ancestor as sometime after Lucy, less than 3½ million years ago. But Johanson is not, of course, suggesting that the split he dates so neatly to almost the same time that the molecules give for the origin of the human line is the split that produces man and the African apes. Why not? Essentially because 'everyone knows' that the man–ape split took place at least 20 million years ago.

Let's look a little more closely at the evidence. The australopithecines that descended from a Lucy-like ancestor were around until about a million years ago, according to the fossils that have been unearthed so far. These cousins of ours, living in the same part of the world as early man but not members of man's species, walked upright, and came in two varieties. The larger *Australopithecus robustus* stood about 5 feet tall, while the slender *Australopithecus africanus* was a foot shorter and probably weighed around 30 kg. The two forms are often referred to as the robust and gracile australopithecines respectively. 'For about a million years,' Johanson tells us, 'they appear to have walked side by side' with *Homo*, but 'by one million [years ago] there were no australopithecines left. They had all become extinct.'[13]

Extinction is one of the few biological absolutes, but it is one that is hard to be absolutely certain of. Is Johanson's assumption that the australopithecines became extinct justified? Gaps of a million years in the fossil record are far from uncommon, and one of Johanson's great hopes for renewed fieldwork in Ethiopia is that he will find fossils from the period between 3 and 2 million years ago. This gap of a million years he describes as a 'black hole', a 'pall of ignorance' in palaeo-anthropology.[14] But supposing the two australopithecines did not go extinct; suppose we say instead that there is a 'pall of ignorance' over their subsequent evolution. What might they have evolved into? The best basis we have for making a guess comes back to those teeth. Alan Walker, who works in Baltimore at Johns Hopkins University, has been using a new electron microscope to study the minuscule scratches, grooves and pits on the teeth of different living species and fossils. The markings on a tooth are created by the diet, and by relating the pattern of the markings to the food of present-day species, Walker can work out the relationship between diet and microscopic tooth wear. Then, when he looks at the wear on a fossil tooth, he can make a very good guess about the sorts of food that fossil tooth had to cope with. Johanson finds Walker's work 'marvellous', and describes the startling implications that emerge from it: 'The polishing effect he [Walker] finds on the teeth of robust australopithecines and modern chimpanzees indicates that australopithecines, like chimps, were fruit eaters. That news comes as a surprise. Everything we have learned about australopithecines – that they were ground-dwelling, bipedal, savannah-frequenting creatures – suggests that they were omnivores. . . . If they were primarily fruit eaters, as Walker's examination of their teeth suggests they were, then our picture of them, and of the evolutionary path they took, is wrong.'[15]

Perhaps our armchair speculations can provide Johanson with a solution to his problem. For, if we accept Walker's evidence at face value, the split from a Lucy-like ancestor into two or more types of hominid exactly fits our picture of one side of the family giving up the difficult business of becoming human and settling down to a sybaritic life of ease beneath the trees, living off the ample supply of fruits and obtaining a free lunch. But a well-known saying reminds us that 'there's no such thing as a free lunch'; everything has its price, and the price the australopithecines paid for their lunch was that they gave up the chance to become human. (Whether this is so desirable a goal, we are unqualified to judge.) They became more ape-like, a term that may annoy some palaeontologists but is nevertheless abundantly clear. Could they, indeed, have become so ape-like that they are represented today by their living descendants, the chimpanzee and gorilla? The fit with lifestyle, and with the dates of the molecular clock, is so impressive that the idea deserves to be taken seriously. There are problems, of course — for example the one we mentioned about hip structure — but the reversion to a former type could have been perfect. And, as Johanson made clear, there are even bigger problems, it would seem, with the conventional picture of *Australopithecus*.

The mystery is one that fans of Sherlock Holmes would appreciate. On the one hand, we have two related species, large and small variations on a theme, that split from a common ancestor with the human line around 4 million years ago. They were fruit eaters, and can be traced in the fossil record to one million years ago. On the other hand, we have two hairy apes living in Africa today, one large and one small, both close relatives of the human line, from which they diverged about 4 million years ago. Both of them eat fruit, but palaeontologists will tell you that no recent ancestors of either chimp or gorilla have been found in the fossil beds. So, we have two living species, without known ancestors, and we have two fossil species, without known descendants. They all split from the human line at about the same time, and they all eat fruit. As Johanson, quoting Euclid with approval, says of his fossil teeth, 'things that are equal to the same thing are equal to each other'.[16] And as Holmes admonished the good Dr Watson, 'How many times have I said to you that whenever you have eliminated the impossible, whatever remains, *however improbable*, must be the truth?'.[17]

Ockham's razor, the maximum parsimony principle, surely tells us that the simplest assumption, until proof to the contrary turns up, is that the two hominid species that split from our line 4 million years ago, and left fossils until one million years ago, are the ancestors of the two species alive today that are most closely related to man. Once again we ask, why not? A final irony is that Johanson, unlike us, may have literally held the answer to this part of the monkey puzzle in his own hands. His colleague Tim White spent painstaking months using all his skills and all the fragments at his disposal to create a composite skull of *Australopithecus afarensis*, Lucy's species. Eventually, after several setbacks, White finished the job and showed the skull to Johanson. 'It looked,' Johanson says, 'very much like a small female gorilla.'[18]

Is there any other fossil evidence to support our speculations? In his book *The Making of Mankind*, Richard Leakey mentions the work of Ralph Holloway, an anthropologist at Columbia University in New York City. Holloway has perfected the art of making a cast of the brain by swirling a latex solution inside a hollow skull. He can use the method with fossils, and deduces the nature of the brains of our ancestors by studying the interiors of their fossilised brain-cases. 'The basic shape of the human brain,' says Holloway, 'is clearly evident in the hominids at least two million years ago.'[19] This, says Leakey, was a surprising discovery to the palaeontologists, 'as the size of the australopithecine brain was not dramatically different from that of a chimp or gorilla brain'.[20] The way we see it, this is no surprise at all.

We do not say dogmatically that the modern chimp and gorilla are the descendants of *A. africanus* and *A. robustus*; we do say that this possibility resolves so many aspects of the monkey puzzle that it should at least be looked at seriously by the experts and not dismissed out of hand. Johanson's experience with Lucy and the australopithecine teeth points up the need to keep an open mind on such issues until the proof, one way or the other, really is conclusive. Were we real palaeontologists ourselves, the one real issue that we might take up with Johanson is his naming of Lucy and her type; rather than *Australopithecus afarensis*, we might prefer to see them labelled *Homo afarensis* in deference to their undoubted ability to walk perfectly upright. Such a naming would also indicate clearly the possibility that the australopithecines, and perhaps the chimp and gorilla, diverged from our own lineage after upright walking had been invented. But this is a human chauvinist view; no doubt the chimps and gorillas might feel that their two species (despite being different names) should hold sway over our single human species. Johanson's classification, whatever else it may be, is democratically sound.

We may suggest that the apes have evolved 'from' man, but this should be understood in the same way as the common view that 'man evolved from the apes'. Not that the common ancestor was a fully developed human, but that it was more human-like than ape-like. And as a final caveat, we should stress again that there is no real sense in which any species on Earth is more evolved than any other. Those that are here are an evolutionary success, and those that are no longer here were a success in their time. Just because we came later on to the scene than, say, the dinosaurs does not mean that we are superior to them in any way. And that, perhaps, is just as well, for what we are suggesting is that the chimpanzee and gorilla are later arrivals on the evolutionary scene than man himself, and might be desceneded from proto-human stock. They may mark a successful re-adaptation of the human line to the return of lush conditions in Africa, while we represent an older primate form, adapted to harsher conditions. Indeed, we might well have died out altogether but for a further series of climatic changes that put adaptability at a premium.

So why didn't the human line die out? Nature seems to be extremely conservative, and evolutionary types do not vanish for no reason. Indeed, some very ancient forms of single-celled life, essentially identical to the ancestors of all other life on Earth, still proliferate on the planet we like to think of as our own. It is no stranger that they should remain, while evolution explores multicellular possibilities, than that we should have remained while our brothers returned to the soft life. And when conditions changed again, we were ready in the wings to take advantage of them. Today, we put other apes in cages and feel superior; we were given the opportunity to do so only by a bizarre series of geographical and astronomical accidents that produced a pattern, quite possibly unique in the history of the Earth, of repeating Ice Ages.

Eight

PEOPLE OF THE ICE

The landscape of Europe and North America is littered with traces of a major catastrophe. Rocks are scratched and marked as if by a great scouring; a blanket of sediment covers the ground; massive boulders lie, sometimes in jumbled heaps, many miles from the rock formation they belong to, and equally far from any great rivers whose floodwaters might have carried them to their present resting place. No geologist could ignore such phenomena, and for centuries these traces of a great natural disaster were interpreted as evidence of the Biblical flood. It was only in the 19th century that a few scientists began to question this explanation, doubting that a flood, even of Biblical proportions, could really shift huge boulders across hundreds of miles. And it was only in 1837 that Louis Agassiz, who at the age of 30 was already President of the Swiss Society of Natural Sciences, astonished his colleagues by addressing their annual meeting not with a paper discussing the origin of fossil fishes, as they had expected, but with a discourse on the origin of the scratched and polished boulders of the nearby Jura mountains. Agassiz claimed that these boulders, called erratics, could only have reached their present locations if they had been carried by ice, which would also have caused the observed damage. Not a flood covering the face of the Earth, but a great sheet of ice stretching right across Europe. That was the image Agassiz conjured up for his surprised colleagues, initiating a debate that raged for a quarter of a century before the Ice Age theory finally gained acceptance.[1]

Agassiz was not the first to make the connection between erratic boulders and past glaciation, but he was the first influential scientist actively to promote the idea. A Swiss minister of the cloth, Bernhard Friedrich Kuhn, had interpreted the evidence correctly as early as 1787, and in the 1790s the Scot James Hutton had visited the Jura and had realised that glaciers had been at work there. Hutton was a medical doctor who gave up medicine for farming at 24 and retired from his farm at 42. He is widely regarded as the father of modern geology, and was a profound influence on the succeeding generation of scientists, but his writing was obscure, and neither Hutton, nor Kuhn, nor indeed any of the others who independently reached the same conclusions about the erratic boulders of the Jura had the aggressive spirit needed to take the Ice Age hypothesis forward, testing it in the arena of scientific debate until it had been tempered into an established theory. A Swiss naturalist of the time, Jean de Charpentier, put the idea onto a firm scientific footing in the early 1830s, but the scientific establishment held firm to traditional ideas, bolstered by faith in the literal word of the Bible. Young Agassiz had met de Charpentier while still a schoolboy in Lausanne; but he too initially rejected the idea, despite his respect for the older man. Eventually, however, Agassiz was persuaded by the weight of the evidence, and he became the forceful front man and advocate that the Ice Age theory needed.

Like that other great scientific theory of the mid 19th century, the Ice Age theory met with a hostile reception at first. Alexander von Humboldt, naturalist and explorer and remembered today for the ocean current that bears his name, wrote to Agassiz and urged him to give up his obsession with glaciers and concentrate on fossil fish, since 'in doing so you will render a greater service to positive geology, than by these general considerations (a little icy besides) on the revolutions of the primitive world, considerations which, as you well know, convince only those who give them birth'.[2]

Fortunately, Agassiz did not give up, and did eventually convince others. The detailed interpretation that Agassiz put on the evidence of past glaciation in Europe certainly was sensational. Catastrophism was the accepted scientific view of world history at that time, the idea being that great natural disasters had repeatedly wiped the slate of life clean and allowed for the creation of new forms of life after each disaster to repopulate the Earth. Baron Cuvier, the French academician who did so much to rescue the Church from the increasingly embarrassing finds of the fossil-hunteres, was a leading light in the catastrophist movement, and after his death his colleagues in the Academy calculated that there had been no fewer than 27 separate Acts of Creation, all but the last wiped out by a great catastrophe. (William Smith, the English canal engineer and amateur geologist, who as a result of his work digging ditches through the countryside gave rise to the branch of modern geography called stratigraphy, later upped the number of strata, and hence creations, to 32.) It was against this background that Agassiz saw the great ice sheets of a past geological epoch as the mechanism of a great catastrophe at work.

'The development of these huge ice sheets must have led to the destruction of all organic life on the Earth's surface. The ground of Europe, previously occupied with tropical vegetation and inhabited by herds of great elephants, enormous hippopotami, and gigantic carnivora became suddenly buried under a vast expanse of ice covering plains, lakes, seas and plateaus alike. The silence of death followed . . . springs dried up, streams ceased to flow, and sunrays rising over that frozen shore . . . were met only by the whistling of northern winds and the rumbling of the crevasses as they opened across that huge ocean of ice.'[3]

The Ice Age theory needed a spokesman it is true, but ironically the very extravagance of Agassiz' claims played a part in the way that more sober academics dismissed the theory. Agassiz spoke of ice sheets extending to the Mediterranean and South America, claims so outrageous that many geologists initially threw out the entire Ice Age hypothesis. But erratics and polished pebbles, smoothed by the action of ice, continued to be discovered not only in Europe but also in North America, where in 1839 palaeontologist Timothy Conrad was the first to publish a paper suggesting that polished rock surfaces in New York State were the result of former glaciation. By the middle of the 1860s, despite Agassiz' initial enthusiastic overkill, the Ice Age theory was well established, although critics remained to the end of the century. In the second half of the 19th century the sensational scientific debate was over Darwin's theory of evolution and natural selection, not Agassiz' theory of glaciation. But 20th century science was to show not only that Agassiz' claims were not all as outrageous as they had earlier seemed, but also that the climatic history of the Earth, and the ebb and flow of great Ice Ages, was indeed intimately linked with the story of evolution.

In Agassiz' day, the talk was all of *the* Ice Age, a unique catastrophe that had struck the Earth in the not-too-distant past. But once the marks of glaciation and other traces in the sedimentary record had been accepted as proof of past ice movements it soon became clear that the glaciers had marched back and forth across Europe and America not once but several times. By the beginning of the 20th century the advance and retreat of the ice sheets had been traced in rocks dating back more than half a million years, and the evidence was thought to indicate four major Ice Ages over the past 600,000 years, each lasting for a few thousand years and separated by much longer warm intervals.

Geological advances in the second half of the 20th century have, however, turned this picture completely on its head, just as they have transformed just about everything else in what are now called the Earth sciences. It hasn't simply been a question of better techniques for dating and analysing very old rocks, or for interpreting the more recent material, though those have been important; it has also been a series of major conceptual break-throughs. One consequence of the revolution in Earth sciences is that the theory of continental drift — today dubbed plate tectonics — is well established as a powerful explanation of major changes in the condition of the surface of the Earth throughout its past. Another key development was the realisation that the 'Great Ice Age' was neither a unique event nor even one of a series of four or five smaller Ice Ages restricted to the past few hundred thousand years. Rather, it seems that at intervals of a couple of hundred million years or so, throughout the past thousand million years or more of Earth history, the planet has been subject to Ice Epochs millions of years long. In the middle of the Permian era, for example, 250 million years ago, there was a great Ice Epoch that lasted for 20 million years, leaving its traces in the record of rocks throughout that time. Within an Ice Epoch the glaciers ebb and flow to be sure — but full Ice Age conditions are the dominant norm, with warm interglacials the short-lived exception rather than the rule.

After the end of the Permian Ice Epoch, the world experienced 200 million years of warmth. During that long interval the dinosaurs first flourished and then vanished, leaving vacant the ecological niches that were to be filled by the adapting mammals. The primate line arose during the conditions of global warmth that have been, on the longest timescale, normal during the Earth's history. But during most of the subsequent evolution of that line the Earth experienced a long slow cooling that began about 55 million years ago. About 10 million years ago glaciers returned to the face of the Earth, at first in a small way on the mountains of Alaska, but much more dramatically in Antarctica, where the massive Antarctic ice sheet quickly grew to about half its present size. Some 5 million years ago the Antarctic ice sheet may well have been even larger than it is today, and by about 3 million years ago the first glaciers of the modern Ice Epoch had appeared on the continental land masses around the North Atlantic Ocean.

(189)

For 3 million years, the Earth has been in the grip of a full Ice Epoch, with frozen ice caps at both poles. Sometimes the ice advances; much less often it retreats a little, and conditions in Europe and North America (and many other places) become more equable and pleasant. We are living in just such a pleasant warm interlude, an interglacial. But we are also living in a great Ice Epoch, and it is only a matter of time — a very short time — before the glaciers return in full strength to initiate the next phase of full Ice Age conditions. We can imagine how Agassiz would have delighted in that piece of information, but we cannot guess how he would respond to the full story.

Professor Hubert Lamb, one of the grand old men of climatology, has summed up the evidence for the great cooling in his *magnum opus, Climate: Present, Past and Future*. Interpreting the evidence with due caution, he says:

'At some stage during the cooling after the mid Miocene [the geological epoch from 25 to 7 million years ago] great ice sheets did come into existence in Antarctica, certainly by 7 million years ago. It seems probable that the first continent-wide ice sheet formed on the central plateau of East Antarctica . . . It was probably only after this great ice sheet was there that the climate could become severe enough for the glaciers in the high mountains of West Antarctica to produce thickening and coalescing ice-shelves on the deep ocean among the islands of what was then an archipelago, and for these ice shelves to become grounded and ultimately fill the ocean deep in that area.'[4]

Coming closer to the present, Lamb comments that in the period before one million years ago there were several glaciations of varying extent in Argentina and that ice sheets extended into the Santa Cruz valley at 50 degrees south, slightly further north than the Falkland Islands. The very earliest of these Ice Ages, dated by the potassium/argon technique, took place just over 3.6 million years ago, so that by that time the oceans must already have cooled a great deal, and the associated worldwide climatic changes had begun. That date — 3.6 million years ago — is a key date in the development of the present Ice Epoch, and it coincides with the beginning of a key period in human evolution. It was just before then that the hominid line emerged from the three-way split of man, chimp and gorilla; it is

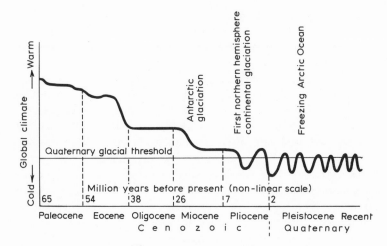

The downward slide of global temperature since the Paleocene, 65 million years ago. Note that the horizontal time scale is not linear, so that more recent variations, largely the result of the Milankovich process, show up more clearly. (Modified from J. Andrews, in *Winters of the World,* ed. B. S. John, David & Charles, 1979; schematic only.)

also at that time that we have the first incontrovertible proof of a hominid walking upright, not just the footprints in the lava but also Lucy's knee. Coincidence? It hardly seems likely, for evolutionary events in East Africa were certainly influenced by the same pattern of climatic changes that caused those far off events in Antarctica: those events mark the beginning of our own era, the Quaternary, so we ought to look at them in a little detail.

It is the onset of the severe phase of the latest Ice Epoch that marks the boundary between the Tertiary and Quaternary periods of geological time, although the boundary is more commonly referred to by the names of the sub-units, the geological epochs, on either side. This gives us the term Pliocene-Pleistocene boundary, which is synonymous with Tertiary-Quaternary. Geologists measure time not in terms of an absolute calendar or clock but in relation to certain events, literally landmark events, that have left their traces on rocks around the globe. The Pleistocene epoch, and the Quaternary period, start, by definition, with the start of the severe phase of the present Ice Epoch, but there has been considerable debate over when exactly all that happened. To some extent, of course, the debate is unnecessary and the definition arbitrary, because the world did not plunge into the Ice Epoch overnight, even on a geological timescale. It slid into it inexorably, and it is we humans who draw lines on geological charts and say 'this side Tertiary, that side Quarternary'. Nevertheless, it is convenient to stick to round numbers in this case and set the boundary, the beginning of the Quaternary, at 3 million years ago. Within the Quaternary we recognise two epochs, the Pleistocene, which takes us up to 11,000 years ago, and the Recent, in which we are living. But the boundary between Pleistocene and Recent is entirely an invention of the human mind. It is defined to mark the end of the most recent phase of glaciation, the start of the present interglacial — but we now know that the present Ice Epoch is far from over, and that the interglacial that supports us will be followed by further waves of ice. It looks special to us only because we are alive now to look at it. On the other hand, as we shall see later in this chapter, it may be that we are alive to name the geological periods solely as a result of the events that produced the previous boundary between epochs, the Pliocene-Pleistocene boundary. To see why, we must go back to that boundary, and beyond, to place human evolution, and life itself, in perspective on the changing face of the Earth.

All of these changes, the dramatic onset and equally dramatic ending of the Ice Epochs, are neatly explained within the framework of plate tectonics, the modern theory of continental drift. In simple terms, the surface of the Earth is not a fixed and unchanging crust, with present-day geography established once and for all, but a kind of spherical jigsaw, in which the pieces of the puzzle, the continents, are free, on a timescale of millions of years, to move around the globe and form new patterns.[5] Ice accumulates on land only when that land is located at high latitudes, at or near one of the poles. When there is no land near the poles the land itself is kept warm simply because it is nearer the equator, and the polar regions are kept warm by a continual influx of warm ocean currents, which carry their burden of solar heat right up to the highest latitudes.

The great Permian Ice Epoch fits this picture neatly. Two hundred and fifty million years ago, according to modern geophysical reconstructions of the continents, the whole of the Earth's landmass was grouped in a single vast supercontinent called Pangaea. The centre of Pangaea was on the equator, but it was such a vast continent that its southern boundary extended right down across the South Pole. The regions of land that we know today as Brazil, Argentina, South Africa, Antartica, Australia, and even India were, 250 million years ago, the regions of Pangaea at high southern latitudes. They are also the regions of the world where today we find evidence of glaciation on rocks of Permian age.

Fascinating though such reconstructions undoubtedly are, their direct relevance to the story of human evolution is minimal. Anything that happened before the death of the dinosaurs 65 million years ago is of only tangential relevance to the evolutionary pressures that made a man out of a monkey. But the latest Ice Epoch — the one in which we are living — and the events that led up to it are of crucial significance in the story of man's emergence.

The Ice Epoch of the Quaternary in general, and the past million years especially, has been marked by variations in which the dominant Ice Age conditions are temporarily displaced by warmer interglacials with a very regular repeating rhythm. Even when the Ice Age remains, the temperatures fluctuate on the same rhythm. This pattern has been picked out by the analysis of sediment cores from the deep ocean basins, focussing on the measurement of the isotope oxygen-18. Microscopic sea-creatures use calcium, oxygen and carbon to build their chalky shells, and the amount of oxygen-18 available depends in an accurately determined way on the temperature of the ocean that the creatures lived in. So the shells of these micro-organisms contain a fossil thermometer, waiting to be read. In fact it was only in the mid-1970s that scientists were able to read the thermometer with definitive accuracy.[6] They showed that in the past million years there have been 10 glacial maxima — Ice Ages — as severe, or more so, than the one that peaked 20,000 years ago and that Agassiz thought of as *the* Ice Age. In the million years before that, from 1 to 2 million years ago, the cold periods, the maxima of glaciation, were rather less intense, and the pattern of Ice Ages was dominated by a rhythm of some 40,000 years. Since a million years ago, however, the pattern has been a great deal more complex.

Detailed analyses of a sediment core spanning the past half million years show, in addition to the 40,000-year cycle, a dominant period of 100,000 years and another strong effect with a period of 24,000 years.[7] The overall effect of the three cycles running together is that an Ice Age lasts for about 100,000 years, during which time the amount of ice cover fluctuates according to the other two rhythms, and then ends abruptly, ushering in an interglacial that lasts for about 10,000 years before the ice returns. Astonishingly, although it is only in the past decade that we have obtained the proof of this repeating pattern of ice ages, the theory that explains it existed long ago, when the existence of even a single Ice Age was still being hotly debated. That theory, now firmly founded on impeccable data, depends on changes in the geometry of the Earth's orbit around the Sun; it goes back to some speculations of the famous astronomer John Herschel in the 1830s, and was put into formal shape by French mathematician J. F. Adhémar in 1842. But it was a Scot, James Croll, who in the 1860s developed Adhémar's ideas fully, at the same time as the scientific establishment was beginning to accept the proof of a past Ice Age.

Croll calculated the details of the way in which the Earth's wobble in space — its precession — changes the amount of heat arriving at different latitudes in different seasons. This was a remarkable enough achievement for the mid 19th century, involving laborious pencil and paper arithmetical calculation of the gravitational interactions not just between the Earth and Sun but also between the Earth and the other planets. Even more remarkable is the fact that at the time he was labouring over these calculations Croll was not exactly a member of the inner circles of academe; he was, in fact, the janitor at the Andersonian College and Museum in Glasgow. The story, alas, is not quite the one of unsung genius that this image conjures up, for Croll was a janitor in the same sense that Albert Einstein was a patent clerk: it provided a quiet place to get on with the real work of life.

Croll, born in 1821, had a lifetime interest in science and philosophy, but came from a family too poor to afford the luxury of a university education; he tried employment as a millwright and carpenter, always reading and studying as widely as possible, and then turned to a succession of physically less demanding jobs because of the deteriorating condition of an injured elbow. A shop and a hotel both failed (at least partly because Croll devoted so much attention to academic, rather than business, books), and he even tried a spell as an insurance salesman. During a period of unemployment he completed and saw published a book, *The Philosophy of Theism*, which was not only favourably reviewed but actually made a small profit, and in 1859 he took up the janitoring job. The pay was modest, but Croll commented, 'I have never been in any place as congenial',[8] for now he had access to a first-class scientific library. Studying more intensively than ever (one wonders how much janitoring got done), Croll began to publish scientific papers, and touched on Ice Ages for the first time in 1864, in a contribution to the *Philosophical Magazine*. By 1867 the impact of his ideas had become so great that he was offered, and accepted, a post with the Geological Survey of Scotland. In 1876, a year after the publication of his book *Climate and Time*, Croll was elected a Fellow of the Royal Society. The sometime millwright and carpenter, failed hotelier, unemployed insurance salesman and janitor had truly made his mark on science.

Croll got there first, but the person whose name is now permanently linked with the astronomical theory of Ice Ages is Milutin Milankovich, a Yugoslav astronomer who picked up the threads of Croll's work in the 20th century and by 1938 had published a complete theory of Ice Ages, pinpointing changes in the amount of summertime heat as the prime movers of glacial fluctuations. Milankovich also calculated 'predictions' of the extent of ice cover during the past million years, but the geological evidence was too sparse to provide accurate comparison with Milankovich's predictions. The theory — now called the Milankovich Model — could not then be tested; when geological practice caught up with Milankovich's theory, in the 1970s, the fit between predicted and observed patterns proved to be all but exact.

Perhaps the most remarkable thing about all this is that it seems most unlikely that the Milankovich Model would be of any use in explaining the Ice Age rhythms of Ice Epochs past, the Permian glaciation for example. It seems that the Earth today, and for the past few million years, is, and has been, in an unusually sensitive state, fine-tuned to respond to the astronomical rhythms of the Milankovich Model. And the fine tuning is related to the geography of the globe today.

The usual state of the Earth is to have a variety of continents scattered about its surface with a free passage for ocean waters to carry the warmth of the tropics up to the cold polar seas. Just occasionally, roughly every 250 million years or so, a continent drifts across one pole of the Earth, blocking off the flow of warm water. Glaciers build up on the surface of the polar continent, and as long as it stays in this unusual position (in round terms for between 5 and 10 million years) the glaciers remain and affect the entire Earth's climate. Eventually the continent drifts off again, freeing the polar region for the benign ocean waters to return and losing its ice as it slowly moves to lower latitudes. It just so happens that Antarctica today provides a text-book example of this kind of rare glaciated continent. Before it drifted across the South Pole, Antarctica was a tropical continent swathed in forest, as the presence of coal deposits laid down from the forest remains reveals.

Antarctica, then, is a rare ice-cap continent. But there is another, even rarer, way to get an ice-covered pole on a planet like ours. If several continents happen to move into positions *around* the polar sea, they can cut off the flow of warm tropical water to the polar ocean even though none of them sits squarely on the pole itself. When that happens the polar sea can freeze, producing an ice cap floating on the water, and the continents around the pole become so cold that the ice can stretch down across them. Such a situation is obviously much rarer than one continent drifting across the pole, and nor can it last as long, since it is easy for a gap to open up and allow warm water back into the polar sea. But again, it just so happens that the North Pole of our planet today provides an example of this type of glaciation, with an ice-covered Arctic Ocean almost completely surrounded by land masses.

The combination of *both* kinds of polar glaciation on the same planet at the same time must be very rare indeed; there is no way we could put a figure on it, but we feel certain that it is a unique event in the 4½ thousand million year life of our planet. And this unique combination, not just two glaciated poles but one landlocked frozen polar ocean and one ice-covered polar continent, is what has made the present Ice Epoch so special, so sensitive to the Milankovich rhythms that send the ice ebbing and flowing across the globe.

If the Earth were positioned in space so that a line joining the centre of the Sun and the centre of the Earth made a perfect right angle with the line joining the North and South poles of the planet, and if the Earth were a perfect sphere, then there would be no seasonal variations in weather and no rhythmic pattern in the Ice Ages of the past few million years. The cycles in and out of a full Ice Age that have been so common during man's evolution are very much like seasonal variations but on a grand scale, and each year as the seasonal cycles bring winter we get a glimpse of Ice Age conditions. Because the Earth is tilted relative to the imaginary line through the poles, but maintains the same orientation in space as it circles the Sun, each polar region points first towards the Sun and then away from it. When a hemisphere is pointing towards the Sun, the Sun, as viewed from that hemisphere, rises high in the sky; the days are long compared to the nights, and it is summer. In a hemisphere tilted away from the Sun the nights are longer than the days, and even at noon the Sun is low in the sky; it is winter. When the Sun is low in the sky its rays strike the Earth a glancing blow, and what energy they contain is spread over a larger area; that is why winter is cold. In summer the Sun is high in the sky, its energy is concentrated on a smaller area and it is hot. The other two seasons, of course, mark the intermediate stages when the poles are pointing neither towards nor away from the Sun. It is tilt, then, that makes the Earth seasonal. The *amount* of the all important tilt, however, changes as the Earth wobbles through space, and that change alters the balance of the seasons, so that sometimes winters are very cold and summers very hot, while at other times the seasons are more similar.

The main reason for this wobble is that the Earth is not spherical; it is slightly pear-shaped and also bulges at the equator. The gravitational pull of the Sun and Moon (and of the other planets in the Solar System) tugs at the bulging equatorial region of the Earth and sets up a wobble, exactly as the gravity of the Earth tugs at a bulging spinning top and causes it to wobble as it spins. A top wobbles every two or three seconds, but in the case of the Earth the wobble repeats over a cycle that varies between 23,000 and 26,000 years long, the precession of the equinoxes. And the precession means that the poles point at different stars at different times. Right now, the north star Polaris is over the North Pole; it stays still while all the other stars appear to circle overhead. Three thousand years ago the north star was Thuban in the constellation Draco, and in 12,000 years' time it will be Vega. A rather longer cycle, about 40,000 years long, changes the actual tilt of the Earth relative to the Sun, and the longest cycle of importance to these Ice Age rhythms involves changes in the shape of our orbit around the Sun. Over a period of about 100,000 years the orbit stretches from nearly circular, to more elliptical, and back again. In an elliptical orbit the Earth is closer to the Sun — and therefore hotter — at some times of the year than at others.

All of these effects — wobble, tilt and stretch — combine to change the amount of heat arriving at different latitudes in different seasons, but there is never any change in the total amount of heat that the Earth receives from the Sun in the course of one full year, one orbit around the Sun. The argument over how, or indeed whether, these seasonal changes could account for the advance and retreat of the ice sheets has been a long one, which it would not be appropriate to discuss in detail here. But it is now widely agreed that within the present Ice Epoch the pattern of full Ice Ages and interglacials has followed very closely the seasonal variations predicted by the Milankovich Model. Clinching evidence came, as we've seen, in the mid 1970s, when a team from the Lamont-Doherty Geological Observatory in New York published its analysis of isotope variations in deep sea sediments.[9] But at first their results seemed far too good to be true. The influence of the astronomical rhythms, especially at 23-26,000 and 100,000 years, is far stronger than anyone had predicted. Indeed, Croll himself concentrated only on the 40,000 year rhythm, which we now know dominated from 1 to 2 million years ago but has since played a lesser role than the other two Milankovich cycles. Galvanised into action by this discovery the theorists have been hard at work in the past few years, and can now explain most of the observed details of the Ice Age rhythms of the recent Ice Epoch.

The key is that the normal condition during an Ice Epoch is that of a full Ice Age; it is interglacials that are rare. This is the reverse of normal conditions during the whole history of the Earth; then warm periods are the norm with Ice Epochs exceptional. So the problem to be solved is not the one that puzzled our Victorian forebears, of why the Earth should sometimes be covered in ice. Rather, the puzzle is why, during an Ice Epoch, there should ever be interglacials in which the world is *not* covered with ice. But it is more natural for us to look first at the way the present interglacial will end.

To make a great ice sheet in the northern hemisphere, the Earth must spread a covering of snow across the *land* that surrounds the polar regions; the Arctic Ocean, of course, is frozen already. Snow falls every winter around the polar regions; the problem, if you like, is that it melts in summer. So an Ice Age proper will develop when northern *summers* are cold, and the snow that falls in winter stays until the next winter. Once that begins to happen, snow layers build up year by year into great ice sheets — not grinding down inexorably from the pole but growing from the ground up as this year's snow falls on top of last year's, squeezing it down to make ice. This surprising new story explains why some climatologists now refer to Ice Age conditions developing in a single year — *one* cold summer could just about tip the present balance irrevocably into Ice Age conditions, because a layer of snow that persisted through the summer would help to keep high latitudes cool by reflecting solar heat off its shiny white surface.

What about the southern hemisphere? There, with a frozen polar continent surrounded by open sea, the requirements are exactly reversed. Snow falling on the sea simply melts. The way to make the southern ice cover grow is to have very cold *winters*, freezing ever increasing amounts of the ocean water. Once that happens, the white ice sheets will both reflect away the Sun's heat and provide a base for the snow to settle on.

So to make a great Ice Age affecting both hemispheres today we need cold northern summers at the same time as cold southern winters. That, of course, is exactly what the Milankovich Model predicts, since when it is winter in the south it is summer in the north. The calculations actually pinpoint the key season as cold northern late summer or early autumn, but this is merely a refinement. What matters is that with this understanding the Milankovich Model is firmly established as the best explanation of the pattern of Ice Ages and interglacials stretching back over the past few million years. What is more, all the evidence points to a full Ice Age as the normal condition for the Earth with its present geography. It is only when all three Milankovich cycles are pulling together to drag us out of an Ice Age that we get an interglacial, and even then it lasts for little more than 10,000 years.

Why should the rhythms we see be so strong? In all the theorising so far we have left out one crucial element — the oceans — and the oceans are very important. Indeed, the oceans provide *the* key to climate, storing warmth from the Sun and feeding it into the lower layers of the atmosphere to provide the driving power behind the weather machine. From season to season, and year to year, the effect is a familiar one, smoothing out the extreme temperature differences between winter and summer and making maritime climates like those of England and California so much more pleasant and even tempered than nearby continental climates like those of Poland or Colorado. In summer the oceans help to keep adjacent land cool, and in winter they give up their stored warmth and help to keep the land warm. But on a timescale of Ice Ages and interglacials, the pattern is exactly reversed. Instead of acting as a negative feedback to smooth out the ups and downs, the oceans, the North Atlantic in particular, provide positive feedback, magnifying and enhancing the changes and speeding the swings from one extreme to the other.

First details of this oceanic feedback were published only in 1981 (the Milankovich Model of Ice Ages has a few surprises left), by two more members of the Lamont-Doherty Observatory, William Ruddiman and Andrew McIntyre. [10] Their work is especially important because the few remaining critics of the model tend now to argue that the rhythms measured in the deep-sea cores and the rhythms of variation in the Earth's orbit match only by coincidence, since the astronomical effects are too small to do the job of starting and ending Ice Ages. Maybe; but the combination of astronomical rhythms and oceanic feedback can certainly do the job.

The new picture is clear and simple. Ice advances when the astronomical cycles bring cold summers and warm winters to the northern hemisphere, and sudden deglaciations, like the one which began 15,000 years ago, happen when the summers are warm, even though inevitably the winters are then cold. Leaving out the details, it is now clear that when the Earth moves into a phase of cooler summers, as it is beginning to do today, glaciers are encouraged to spread on land around the North Atlantic, because there is insufficient summer heat to melt them back. But as the Earth cools, there is a lingering warmth in the oceans, which encourages water to evaporate even in the wintertime. The lavish supply of water vapour falls as snow, feeding the glaciers and incipient ice sheets even though the winters are not notably severe. The ocean warmth precipitates the freeze of the land.

By the end of a cycle of advancing ice, the northern Atlantic is largely covered with ice, and kept cool by icebergs that calve from the great glaciers flowing off the land. As the astronomical orbital rhythms move us into a state of warm summers the ice on land begins to melt back and retreat, producing even more icebergs that keep the sea cold and partly covered with ice. So although the winters are now severe, the sea is cold and there is little evaporation. The summer losses in the ice sheets cannot be made up in winter. A cold ocean thus helps the retreat of the ice, and the net result is that when conditions are just right (when all three Milankovich rhythms are working together) the Earth emerges very suddenly from a full Ice Age into an interglacial. A short time later, somewhere between

10 and 13 thousand years, the astronomical patterns have changed and the Earth is plunged even more suddenly back into an Ice Age. Each Ice Age lasts about 100,000 years, but with some ebb and flow of the ice sheets, following the 40,000 and 23,000 year cycles. Each interglacial is a little over 10,000 years long. On the most conservative estimate, the interglacial we are enjoying was in full flush 11,000 years ago.

It is easy enough to see why such a repeating pattern of Ice Ages provided the stimulus for our distant ancestors to move out of the woods and onto the plains, developing intelligence and adaptibility along the way. Geological studies and computer simulations of the weather patterns of an Earth with enlarged polar ice caps tell us that during full Ice Age conditions the rainfall over the planet would be a lot less than it is today. The planet is colder, so less water evaporates from the sea, making fewer clouds to provide any rainfall. A great deal of water that today is involved in the continuing hydrological cycle that brings our rain is, in a full Ice Age, locked up in the great ice sheets that cover the high latitudes. In particular, these simulations and studies show that during a full Ice Age the rainfall in the region of East Africa where man evolved is much reduced. With less rainfall, there is less tropical forest and more open savannah. Some species do well in such circumstances, others do less well and may even die out. Evolution applies its cutting edge, and natural selection may carve out new characteristic adaptations and new species.

In general, the kind of climatic change that we have been talking about ought to have little effect on the woodland species that are very well adapted to their environment. They may be forced to retreat to the heartlands of what were once more extensive forests, and the total number might fall, but their skills in the forest way of life should keep the species safe as long as at least some forest remains for them to live in. Plains dwellers do well, of course, with an extended territory to roam and the opportunity for predators to pick off woodland creatures trying to scrape a living on the edge of the forest. The animal species that would do worst as the climate changed and the forests shrank would be the less than perfectly adapted woodland species.

With less forest available, competition among the trees is fiercer than ever before; those that could not compete would find themselves, perhaps literally, pushed out to forage at the fringes of the forest and even on the grassland. There, they would encounter fast-running predators already well adapted to life on the plains. The prospects, at first sight, could hardly appear worse, and many forest-dwelling species were wiped out as the drought — far more important than the cold of the Ice Age, which hardly affected tropical regions — set in. But one species of not very successful woodland ape was able to take advantage of the changing situation. As we described in the previous chapter, one simple evolutionary step, neoteny, gave our ancestors a vital package of adaptations: it allowed upright stance (useful to spot predators and freeing the hands to carry and manipulate things), delayed brain development (useful for the young to acquire new tricks of survival in a changing world by learning rather than waiting for inheritance), and, as a kind of bonus, provided a hairless skin (which helped our ancestors to keep cool as they ran across the plains, either to catch food or avoid being caught). Neotenous development also enhanced the natural ape curiosity of our ancestors, and the very powerful selection pressure of the new way of life rapidly weeded out those individuals who couldn't walk well, or didn't use their height to watch out for predators, or weren't inquisitive enough to try new kinds of food. For the first time in the history of our planet intelligence became not just a mildly useful asset but a prime survival characteristic for one species. Intelligence and adaptability, more even that the ability to run fast, or see danger a long way off, became the basis of that species' way of life, and have remained so until very recent times.

The dates fit together beautifully. We find evidence of the first major glaciations of the present Ice Epoch in northern hemisphere rocks about 4 million years old, following millions of years of global cooling (and drying) as the continents slid towards their present positions. The molecular clock tells us that the split between man, chimp and gorilla took place at the same time. It is surely reasonable to suppose that this split is a response to changing environmental pressures. Successful species, remember, change very little for millions of years at a time. New species emerge by splitting off from (probably isolated) populations of the old stock when changing conditions provide new variations on the old theme with a selective advantage. Fossil footprints from the lava beds of East Africa demonstrate unequivocally that our ancestors were walking upright 3.8 million years ago. They were walking into an Ice Age — indeed the Ice Age had made them walk — though of course they couldn't know it. And not just any old Ice Age, but an Ice Epoch that was to be dominated by the incessant astronomical rhythms calculated by Croll and Milankovich, rhythms that contrived to produce 3 million years of changing environmental pressure, with no long-term stability. Three million years in which intelligence and adaptability would be at a premium.

The conditions that ushered in the new Ice Epoch should have been disastrous for a great many species, and so it proved. Although less dramatic than the boundary which marks the death of the dinosaurs, between the Cretaceous and the Teritary eras, when half of all genera were wiped out, the present Ice Epoch will provide a clear geological marker for palaeontologists in 50 million years. We cannot say for sure how many species and genera have vanished because we are still far too close to the Quaternary boundary, but we can be reasonably sure that even before man began to wreak his 20th century exterminations a large number of species were exterminated by the very conditions that created man. Mammoth and sabre-tooth tiger are two examples familiar to everyone.

Our Epoch is different, though. Previous Ice Epochs, caused by a continent drifting over a pole, were almost certainly a great deal less sensitive to the fluctuations of the Milankovich rhythms than our own precarious Epoch. The ice came, stayed for a few million years, and vanished as the continent drifted down to warmer latitudes. Species were extinguished as the Ice Age began, and those that survived radiated to fill the gaps. Then, when the ice disappeared, so too did the species that were too tightly adapted to the cold. Again, the survivors radiated to fill the available ecological niches, but all this was taking place on a timescale of millions of years. In our own epoch we have had a repeating pattern of harsh conditions for 100,000 years interspersed with 10,000-year interglacials, a breathing space, if you like, in which species close to extinction would have time, and opportunity, to recuperate. Such a system — a turn of the screw, followed by a slight easing off and then another turn of the screw — could almost have been designed to provide the evolutionary conditions necessary to foster adaptability and intelligence. The Ice Age selects for these capabilities, but exacts a heavy toll so that, numerically, there are few survivors. The interglacial then allows those selected by the previous Ice Age to multiply and spread, creating a bewildering variety of new combinations for the next Ice Age to select from.

Suppose a full Ice Age had set in 4 million years ago and stayed, with East Africa drying out steadily as a result. Would our ancestors have survived on the desert plains? Or would unending harsh conditions over millions of years have led to their extinction? We can only speculate, but it seems likely that those ancestors would, in fact, have died out, as the recent evolutionary experiment of Neanderthal man shows. Neanderthal is interesting particularly because it indicates the sorts of pressures that Ice Ages put on a species like *Homo sapiens*, and the sorts of minor evolutionary changes that must have been a common feature of the past 4 million years. We see the fossils only of species which are successful enough, and last long enough, for their members to be preserved as fossils and then unearthed; the chances of us finding fossils of an evolutionary variation that lasts only a few tens of thousands of years are slim, but the example of Neanderthal serves to emphasise once again how little we can possibly learn from the fossil record about the hominids inhabiting the Earth millions of years ago. There might have been a number of 'Neanderthals' in Africa and in Asia more than a million years ago, but it is most unlikely that we will ever uncover enough fossils to identify these evolutionary experiments.

An Ice Age 100,000 years long gives ample time to begin the process of splitting variants from the common hominid stock, as Neanderthal shows. The least fit, those who breed less successfully, are quietly conquered not by violence but by reproduction. Those who are more fit, who breed more successfully, win out, and slightly different forms of mankind begin to appear. The living is hard, and few survive each Ice Age, but the interglacial gives those survivors the opportunity to spread out and produce new gene combinations to be tested by the next Ice Age. Neanderthal man is the product of only the most recent Ice Age, and it so happens that the Neanderthal line either died out or merged with the central *Homo sapiens sapiens* line. But if we make the modest assumption that in each of the 10 full Ice Ages of the past million years there were similar pressures put on the *Homo* line, then clearly the potential existed for the formation of at least 10 other variations as different from the main line as we are from Neanderthal man. The differences may not be great, but they are important, because in each case the main line ('main' from the viewpoint of ourselves, its descendants) survives and the variation dies out. But if we look at things from the perspective of a million years ago, this talk of a main line is nonsense. We are the descendants of survivors, that is all. It becomes meaningless to argue that one branch is a main line and the others dead ends, and certainly there would be no way to predict which branch will continue and which will die. The line leading to us is unbroken, passing sometimes along an obvious side branch (Lucy, for example, who represented a major departure from the main ape pattern) and at other times staying with the main trunk. We are the end product of repeated selection for intelligence, versatility, adaptability; once you start such an experiment with a species that already possesses a spark of these three qualities it is hard to see how the alternating squeeze of the Ice Ages and nurture of the interglacials could fail to produce something very like us.

The men that flourished during the most recent Ice Age were not failures, as we have remarked before. The remains that have been uncovered portray a well organised society, with hunting more prominent than it had been at any time previously, for the animals that lived, with man, at the edge of the ice were plentiful. Man had time to bury his dead, and even to collect flowers to bedeck the graves, and he also created some of the finest, most vital cave-art known. Neanderthal man, with his heavy brow and narrow chin was just one of the local variations of Ice Age man, and it is unfortunte that earlier interpretations created such an unfavourable press for his kind. They were not shambling, they were not stooped, and they were certainly not stupid. Richard Leakey cites the work of two anatomists, William Straus and A. J. E. Cave, who reassessed the reconstructions of Neanderthal man in the 1950s. Straus and Cave concluded that if Neanderthal man 'could be reincarnated and placed in a New York subway — provided he were bathed, shaved, and dressed in modern clothing — it is doubtful whether he would attract any more attention than some of its other denizens.'[11] (Whether this tells us more about the denizens of the New York subway than about Neanderthal is a moot point.) Neanderthal, a local variant, is best thought of as an extremely successful adaptation to the Ice Ages, dying out, or being reabsorbed, when the ice retreated and truly modern man began to spread.

What of the other races; were they created by the ice?

Before we can discuss the origin of human races we have to be absolutely certain that we understand what we mean by the word race. To the taxonomist, race is something of a weasel word. It denotes a separate geographical population of a species, one which can be distinguished from other races of the same species (though

(205)

often with considerable difficulty); the differences between races are often a reflection of adaptations to geographic differences. In North America, for example, the English house sparrow has spread across the whole continent. Some people have divided it up into a number of races, but in truth there is a simple variation, mostly in size, from more inland northerly areas to more coastal southerly areas. Inland it is colder in winter, and a large body conserves heat better; that is why inland sparrows are bigger. A mating between members of different races is often slightly less fertile than one between members of the same race, because the genetic differences that underlie the racial characteristics are slightly incompatible. In the human case, the whole concept of race is much more problematic.

There are huge differences between humans from different geographic areas, differences that only a bigot would deny. People from cold climates are short and stocky, with flatter faces, while people from hot dry deserts are long and lank. People who spend a lot of time out under the equatorial Sun have very dark skins that protect them from the harmful effects of ultra-violet radiation. People from more northerly climes, and those who live on the equator but in dense forest, have paler skins. And there are plenty of additional differences, in the enzymes, in the consistency of ear wax, and in a host of other strange features.

Anthropologists looking at these adaptations have placed the human groupings into races, and have drawn trees showing how the races are related. In the early days the purpose of this exercise was to demonstrate the biological superiority of the tree-drawer's own race, but now it is done, supposedly, in the pursuit of knowledge. The problem is that the molecules provide a very different tree from outward characteristics. It is notable that there is not a single characteristic that can be used to distinguish races unequivocally. All the features like skin colour, body shape, hair pattern and so on are local variants, and none of them can be used alone to set races apart. The biochemical differences in such things as enzyme structure are also not exclusive to one group or another. And yet we can all recognise racial groupings. We think, though no work has been done on this subject, that it is by the constellation of facial features that we recognise race. And it is the face, squashed in and overshadowed by the massive expansion of the brain, that marks modern man from his ancestors. So could it be that the four major groupings of mankind represent four separate transitions from *erectus* to *sapiens*? We don't know.

Conventional physical anthropology says that race is of very recent origin, 40,000 years at the most, and probably a product of the Ice Age. This completely rules out any possibility that there was more than one 'sapientising' transition. But the molecules suggest a very much more ancient origin for some racial groups (see Epilogue). There is an interesting clash here between politics and science, for the 'recent origin' view of race holds that the differences between peoples are adaptations, so that while all peoples are different they are also equal in those aspects, such as intellect and culture, that are not related to environmental pressures. The 'ancient origin' view sees the truly racial differences as accidents of sapientisation, the particular facial configurations that happened to be prevalent in each of the small groups that made the transition to *Homo sapiens*. These differences are totally unimportant. And the differences that people make so much of, the outward signs of race, are nothing to do with the major groupings but are simply adaptations to local climatic conditions. Evidence either way is not easy to come by, partly because modern scientists haven't looked for it and the older scientists were very selective in what they found. But it is interesting that the front teeth of the *Homo erectus* specimen called Peking man were slightly shovel-shaped, and the front teeth of Andean Indians, believed to have been evolved from an originally oriental stock, show this same feature.[12]

What of the future? The pattern has not come to an end just because one species on Earth has indeed now developed not only intelligence but also civilisation. And it is no coincidence that the whole of human history, the whole of that much-vaunted civilisation, is compressed into one interglacial, a temporary respite from the normal Ice Age conditions that are appropriate for the present arrangement of land masses on our planet. Humankind, and human civilisation, spread so rapidly across the face of the Earth (and now off the Earth) precisely because of the skills and talents that had been whetted by the hardships of the Ice Ages. The improved conditions of the present interglacial that we have the good fortune to find ourselves in have given the intelligent, adaptable species that learned to cope with the rigours of the Ice Age a chance to take over the world.

Will the next full Ice Age, then, bring an end to human civilisation, perhaps within a couple of thousand years? Northern hemisphere summer sunshine, the key indicator, has declined steadily since 11,000 years ago to the point where, other things being equal, there is a real risk of a sudden spread of snow and ice cover, a 'snowblitz' heralding the start of the next Ice Age. Included among the countries that face complete obliteration by glaciers and ice sheets when that occurs are Canada, Ireland, Great Britain, Scandinavia, Nepal and New Zealand. Other countries that would suffer severe glaciation include the United States, the USSR, much of high altitude South America, China and Australia. Most significantly for the near future, many countries would suffer not cold but drought at the beginning of a new Ice Age; these include Mexico, Brazil, the Sahel states of Africa, Pakistan, India, Bangladesh and China. That list of names should strike an ominous chord in the mind of anyone who pays the least attention to world events, since they are all regions where drought has indeed struck in the past few years. That does not mean that the next Ice Age has already begun; it simply confirms the calculations that show that these are regions of the globe that face high risk of drought. It is at least as significant that, this time around, other things are not equal, for the creature created by the previous Ice Ages now has astonishing power over the environment. We may – deliberately or inadvertently – either prevent or delay the onset of the next Ice Age.

The big talking point in climatology today is not the possibility of a new Ice Age but the likelihood of imminent global warming, one result of man's activities. This is the carbon dioxide greenhouse effect. Every tonne of coal we burn, every acre of tropical forest we destroy, contributes a burden of carbon, in the form of carbon dioxide, to the air, and one important property of carbon dioxide is that it holds in heat – infrared radiation – that would otherwise be lost to space. The experts talk of a possible doubling of the natural concentration of carbon dioxide in the next 50 years, bringing an increase in average temperatures of about 3 degrees Celsius, with a bigger increase near the poles and very little in the tropics. Such a climatic change would not be without its attendant problems – not least it might bring a drought to the great food-producing plains of the United States – but it could also herald the beginning of a manmade 'super interglacial', which could hold the next Ice Age at bay for centuries or more, melting more of the polar ice caps than at any time since the present Ice Epoch began. Given another few centuries of progress, our descendants might well have the technological expertise to keep the atmospheric system 'tuned' to prevent the return of the Ice Ages.

In the very long term, by human standards, the Ice Epoch pattern cannot last. Europe and America are separating by a couple of centimetres each year, with the Atlantic widening through the processes of seafloor spreading and continental drift. It will steadily become easier for the warm waters of the Gulf Stream to spread north into the Arctic basin (unless Iceland grows into a continent in its own right, which is not entirely impossible) and thaw the ice cap. By about 50 million years from now, perhaps a great deal sooner, the northern gate will have opened sufficiently to keep the polar sea free of ice, and the sensitive Ice Epoch will have ended.

Looking at the geological record of the Earth's 4½ thousand million year history, it is hard to imagine any recognisably human descendants around to witness such changes. Fifty million years ago our ancestors were very primitive primates. The normal pattern would be for ourselves to be the ancestors (if we are lucky) of species that will have evolved and adapted to the changing conditions and be as different from us as we are from those primitive primate ancestors. Indeed, in 50 million years it is not inconceivable that the human line, or perhaps the whole primate group, or even all the mammals, will have become extinct. But it is also part of the normal pattern that species change very little during periods of environmental stability. We now have the capability to carry that to extremes by making our own environment and adjusting it to our needs. Today we live in air-conditioned buildings, but one day we might air-condition the whole planet. Under those conditions the usual selection pressures are removed. Both of the authors of this book, for example, need artificial lenses to correct the deficiencies of our natural sight and neither of us, for that reason alone, would be likely to be an outstanding success as a hunter or gatherer with no technological aids. Yet the genetic combinations that make for imperfect eyesight cause no problems in a civilised technological society, and people with poor eyesight are no less likely than those with perfect vision to be an evolutionary success — that is, to reproduce and pass on their genes, including those for imperfect eyes. Natural selection depends on differential reproduction, that is, on some people reproducing more than others, and yet there is clear evidence that in some societies, notably the North American cities, there is less differential than ever before. Almost everyone is having the same number of kids, and certainly there is less variation, in terms of breeding success, for selection to operate on. Whether this is so elsewhere remains to be seen (and is probably doubtful for the emerging nations) but the truth is that natural selection is not shaping man today to anything like the extent it did in the past.

In terms of the human species as a whole, the biggest threat — the strongest selection pressure — comes from ourselves. As long as we do not destroy ourselves completely by open conflict, or poison ourselves surreptitiously, there is every reason to suppose that the species will continue on Earth, substantially unchanged, if not for 50 million years then quite possibly for another million. Considering the changes of the past 10,000 years — or even just the past 1,000 years — that is a remarkable prospect. Even if human civilisation lasted for a million years, however, it would still be insignificant compared to the history of a planet on which the present Ice Epoch, 4 million years long, is merely a passing hiccup interrupting the warmer conditions that prevailed for more than a thousand times as long. And yet, it is that hiccup that produced us, and we carry with us the legacy of all those eons of evolutionary history. We are, truly, people of the ice.

Nine
TALK TO THE ANIMALS: BROTHER ESAU

In 1892 an eccentric American, Richard L. Garner, left his home to go on an expedition to the west coast of Africa, to the land that is now Gabon. There, in the equatorial forests, he lived with the Fathers of the Holy Ghost and studied the monkeys around him. He found that the monkeys had a natural language and on his return to America published a book about his work and his discoveries; an unknown critic described the book as 'an odd mishmash of valuable observations, pure inventions, and colourful humbug'.[1] Garner's work is of little scientific importance today, but he inspired the great French writer Jules Verne to produce a little-known novel called *The Great Forest*. In it one Dr Johausen, obviously modelled on Garner, travels to Africa to uncover the language of the apes. Like so many scientists (in fact as well as in fiction) he finds exactly what he sought — speaking monkeys — but there is one essential difference between them and us. The monkeys, unlike men, talk only when necessary. Along the way Verne makes some exceedingly interesting comments about language and what it means to be human, and he ends his story with an ironic twist. Johausen learns the monkey language and rises to become King of the Beasts, with the imposing title to Sa Majéste Msélo-Tala-Tala, but in doing so he loses his own language, his own humanity. In the final analysis, Verne seems to be saying, being human and having language are inextricably inter-woven.

The theme is a common enough one. So close to us are monkeys and apes that we can only wonder why they are not more like us. We erect artificial barriers to keep us apart, and one of those barriers has always been language. Stated most clearly by René Descartes, man alone among the animals has language, and that makes him special. Bertrand Russell also wondered, as he gazed at the rather sad specimens of primatehood dolefully cracking nuts in London Zoo, why they were not human, and what we might tell them to do to make themselves human. Oxford psychologist Dick Passingham has taken up Russell's question in a recent book and ends up siding with Descartes: it is language that makes man who he is.[2] The curious ability of ours to communicate with language has been the barrier of choice since time immemorial, probably since man first realised that there was indeed something special about his way with words. And yet, throughout history, there have been attempts to allow the apes to surmount the barriers and become one with man.

The logic is inescapable: man alone has language; the chimp is very like man; if we could teach a chimp language, he would become something like a hairy man. And it really is nothing new, having its roots at least as far back as the view that language is unique to man.[3] For example, Julian La Mettrie was an 18th century doctor in France who argued persuasively against Descartes' view of language, and provided one of the first clear statements of the talking-chimp lobby.

'Such is the likeness of the structure and function of the ape to ours that I have little doubt that if this animal were properly trained he might at least be taught to pronounce, and consequently to know a language. Then he would no longer be a wild man, but he would be a perfect man, a little gentleman, with as much matter or muscle as we have, for thinking and profiting by his education.'[4]

Jump forward now from La Mettrie's 18th-century France to Nevada in the 1970s. There we find a chimpanzee, a lady rather than a gentleman, walking with her teacher by the side of a lake. 'What is that?' asks her trainer, pointing to a duck. 'Water bird,' replies the chimp, who as it happens does not know the word for duck. The chimp's name is Washoe, and she revitalised the whole question of language in chimps with utterances like this one. But before we examine the performance of these talking apes, and their trainers, we should look back a little at the pioneers who took up La Mettrie's challenge.

People have kept primates as pets since the dawn of history, and none of them ever, to our knowledge, reported hearing the animal mimic the words of its keeper; in this respect a mynah bird seems closer to a man than does a monkey. But at the start of the 20th century some psychologists began a serious attempt to teach apes to talk. In 1909 Lightner Witmer, head of Philadelphia University's Psychological Clinic, reported that he had tested a chimpanzee 'and happily pronounced him a "middle-grade imbecile" '.[5] What is more, the chimp could talk, able to manage the single word 'mama', though Witmer could not deny that the 'm' sound was far better than the vowel, which emerged as a horse whisper. This is the first serious record we have of a talking chimp. The subject of Witmer's study was actually a star performer at the Keiths Theatre in Philadelphia, where he was billed, appropriately enough, as the 'Monkey Who Made A Man Of Himself '. His name was Peter, and he had begun his life of performance by teaching himself to rollerskate while on board ship from England to America. (The skates had been fixed to his feet to slow him down and keep him out of the rigging; they failed miserably.)

William H. Furness III, doctor and chronicler of the head-hunters of Borneo, went one better than his colleague Witmer and began a study of the orang-utan in the wild. He left little record of his expedition, which started early in 1909, but returned with a number of orang-utans and a couple of chimpanzees that he had picked up from a Liverpool animal dealer on his way home. He and Witmer spent long hours trying to educate the apes, and in 1916 Furness reported briefly his scant success. He had managed to teach an orang-utan to say 'papa' and 'cup'. This orang and Peter the talking chimp were the outcome of long hours of painstaking and un-remitting effort. Undaunted, in succeeding years some five couples attempted to build upon these meagre foundations by raising a chimp from a very tender age as a child in their own homes.

In the 1930s Winthrop and Louise Kellogg obtained a chimp, Gua, and brought her up with their son Donald. Gua, sad to relate, never really learned to say any words, though she understood spoken phrases very well and in the early days was far ahead of Donald in this and many other respects. This should not surprise us as we have already seen how much the human infant's growth has been slowed down compared to that of the ape; of course Gua was ahead of Donald at first. The Kelloggs' work was not an overwhelming success, but others had more luck. The acme of ape language, the most loquacious chimp of all, was Viki, raised in their home by Keith and Cathy Hayes. Viki could say 'papa', 'mama', and 'cup', and was just mastering the word 'up' when her training had to stop for reasons we do not know.

The jump to Washoe came when Allen and Beatrice Gardner, psychologists at the University of Nevada, realised that speech was just one form of language (albeit the most prevalent), and it was a form that chimpanzees could hardly be expected to master because their vocal tract is unlike ours and is not able to make the sounds that we can. So immersed were previous experimenters in the effort to teach chimps to talk that they did not take much notice of these fundamental differences. The human vocal tract is unlike that of any other primate in several ways, and these differences are intimately bound up with our ability to speak. We have a short muzzle, and a sharp bend to the tube at the throat, and a nose that can be blocked off at will, all features that enable us to make the complex sounds we call speech. Apes do not have the same structure, so it is hardly surprising that they cannot speak, but does that mean that they can't use language? Perhaps not. The Kelloggs and the Hayes had noted that their chimps were very dextrous, and invented hand gestures with special meanings. The Gardners were watching a silent film of Viki and the Hayes when they saw the potential for using a nonvocal language to communicate with chimps. There was, for the Gardners, a clear connection between the way home-reared chimps used their hands and the sign-language that deaf people use to 'talk'. They obtained a one-year-old chimp in the summer of 1966 and called her Washoe, after the county in Nevada where they lived. The Gardners set Washoe up in a luxurious caravan in their back yard, and proceeded to teach her American Sign Language (Ameslan). Washoe progressed rapidly, learning, it seems, to understand signs and use them too, and flooring her detractors with conversations like the one we reported above. Soon other chimps were being taught language with other nonvocal methods.

Sarah was David Premack's protégée, first in California and then at the University of Philadelphia. Premack taught Sarah a language based on coloured plastic shapes, each of which represented a word or relationship, so that Premack could set up the tokens to ask Sarah a question such as 'What is the colour of apple?' and Sarah could reply by selecting the plastic token representing the word 'red', which was not itself red. Sarah learned quite quickly and was soon doing written tests, in which she would mark the correct answer on a piece of paper, and generally surprising everyone with her abilities. Such was Sarah's skill that she inspired a rather good thriller called *The Poison Oracle*. In this, author Peter Dickinson has a wealthy Arabian sheik who is interested in language pay a young linguist to teach language — Premack style — to a chimpanzee. The chimp is just mastering the names of the people around her when she is the only witness to a murder. The plot hinges on whether the linguist will be able to discover from the chimp who the villain was.

The major advantage of Premack's method is that there can be little argument over which word Sarah used at any time, whereas with sign language it can be very difficult to keep an accurate record of every word a chimp makes. Some might be signs, others simply gestures with no special meaning. Another attempt to overcome the same general problem was made by Duane Rumbaugh and his team at the Delta Regional Primate Center in Atlanta, Georgia.

Rumbaugh's technique takes ape and technology to the limit: his chimps operate a special computer. The chimps, first Lana and then Austin and Sherman, communicated with the computer by pressing special illuminated keys. Each key could contain any one of a number of abstract geometrical patterns, projected onto its surface, and each geometrical pattern, or pictogram, represented a word in the language Rumbaugh devised. The language was called Yerkish, after Robert M. Yerkes, one of America's pioneering prima-tologists. Lana had access to her keyboard all day, and the machine recorded, and acted upon, all her communications. She could ask the machine for food, drink, playmates, movies, and so on, and could also interact with her trainers through the keyboard. The hope was that with the system available 24 hours a day, and with Lana able to obtain what she wanted through language, she would quickly learn what was required. She did, and things seemed to be going very well, with sentences like 'Please machine give Lana orange drink' an everyday occurrence, and occasional widely reported conversations with her trainers about the names of things. In these conversations Lana revealed not only a magnificent command of Yerkish, but showed also that she was a resolute chimpanzee, not to be fooled by some human just because he had access to her chow. Adrian Desmond records this famous interchange:

'Lana needed chow, and Tim [her trainer] deliberately filled the vendor with cabbage. Five times in two minutes she asked for chow, each time Tim insisted that it was in the machine. Realising this could have gone on all night, Lana changed tactics.
Lana: Chow in machine?
Tim: (lying) Yes.
Lana: (exasperated) No chow in machine.

Tim: What in machine?
Lana: Cabbage in machine.
Tim: Yes cabbage in machine.
Lana: (not settling for any monkey business) You move cabbage
out-of machine.'[6]

These three methods — sign language, tokens and pictograms —
produced many astounding results and few critics. They did not
inspire many imitators for the good reason that this sort of work
with apes is time-consuming and costly, and most psychologists
were content to let the Gardners, the Premacks and the Rumbaughs
have all the hassle and to accept their results as unimpeachable. When
Washoe called a duck 'water bird', or a brazil nut 'rock berry', or
when Lucy (another signing chimp) called radishes 'cry hurt food',
many people shouted hurrah and celebrated the fall of another
human bastion. Washoe could invent phrases to name things whose
real name she did not know, using Ameslan symbolically and
proving that chimpanzees can acquire and utilise a language. When
Lana had an argument with trainer Timothy Gill because he refused
to do as she told him and insisted on filling one of her food hoppers
with the wrong sort of food, advocates said that she could detect
prevarication, and if that wasn't linguistic, what was? When Sarah
demonstrated her mastery of cause and effect by linking an apple
and apple slices with a knife, or a wet sponge and a dry sponge with a
glass of water, they said that her performance showed that she had
grasped the mental abstractions needed to join actor and object. A
few killjoys adopted a classic technique of crooked argument and
shifted their ground so as to exclude specifically many of the things
chimps were clearly capable of, ending up with definitions of
language that, taken to the limit, might imply that neither you nor
we truly have language. And the purpose of these philosophical
gymnastics was ever the same, to keep a wall around man that shut
the other animals, even our closest relatives, out. Nobody thought
to question the very ability of the talking chimps. Were they doing
what was claimed of them?

One person who ended up asking just these questions was Herb
Terrace, professor of psychology at Columbia University in New
York, but he didn't start with that in mind. Rather, as a psychologist
trained in the conduct and analysis of experiments on animals, he
could see several loopholes in the work of the Gardners and the

others, and he intended to plug them. Terrace hoped 'that chimpanzees could be taught to use language in a humanlike manner,'[7] and even went so far as to name his animal colleague Nim Chimpsky. (Noam Chomsky is a staunch advocate of the view that language is uniquely human.) 'There were many reasons for starting the project,' Terrace recalled, 'but the most compelling for me was that it might define more clearly what it means to be human. Language and the culture to which it is essential have long been considered the exclusive property of humans. If indeed a chimpanzee could learn to use language in a humanlike manner, the distinction between humans and at least one nonhuman animal would have to be reexamined.'[8] The story of Terrace's efforts with Nim is an intensely moving one, beautifully told in his book *Nim: A Chimpanzee Who Learned Sign Language*, and in one sense it was a great disappointment. When he came to analyse Nim's language in detail, the sort of detail that perhaps ought to have been used in the first place with the Nevada chimps, he found very little evidence that Nim's performance was in fact language.

Take, for example, the famous example of the water bird. Recall that Washoe was walking by a lake when she saw, and apparently named, a duck. There was water present, and there was a bird present. To claim that Washoe described the duck as a water bird, rather than that she named first the water and then the bird (names she was obviously familiar with) is to claim that she was using 'water' as an adjective to modify the word 'bird'. In English, and in Ameslan, adjectives do usually come before the noun they describe, so it is not all that unreasonable to assume that Washoe had created a noun phrase. It was reasonable, but was it true? No, it was not. When Terrace looked at all of Nim's two-word sign combinations he found no tendency for the adjective to come before the noun. The order of the two was random. Nim was not following any rule in his grammar; effectively he had no grammar.

Other objections surfaced too. As a child learns language the average length of its utterances goes up, and as the length increases so too does the complexity of the utterance. Nim's utterances, or strings of signs, also got longer, but they did not get any more complex, and most of the increased length was taken up by repetition. Nim's longest utterance was the sixteen-sign 'Give orange me give eat orange me eat orange give me eat orange give me you'; hardly the same as a young child's 'Would you mind please giving

me some of the orange that you've got in your hand?' Nim inter-
rupted conversations far more often than a young child, so often, in
fact, that they could hardly be thought of as conversations at all. And
when he did reply to signs from his teachers it was most often with
imitations of signs that they had just used, rather than bringing in
new ideas as a child does. Nim was special among talking chimps
because Terrace and his devoted army of helpers had made a special
effort to collect records of all the signs that Nim used with as much
information as possible about context and so on, but the criticisms
that apply to Nim apply equally well to Washoe and the other
chimps that were trained in Ameslan.

As for the other chimp language projects, Terrace (and others
prompted by his revelations about Nim) turned a critical eye on
these studies and found them wanting. Premack's claims and
thoughts on the nature of language, while invariably intellectually
very stimulating, often went beyond the data he had from Sarah and
his other chimps. And Lana and the other Yerkish chimps were not
fundamentally that different from a pigeon in a complicated Skinner
box. If a pigeon learned to peck four coloured keys in one sequence,
say red, green, orange, blue, wherever the colours were, one wouldn't
claim that it was saying 'machine give pigeon grain'. Yet this is
exactly the claim made for Lana. Terrace and his students have
successfully trained a pigeon to do just that, even to the extent of
getting the bird to vary the last colour in the sequence according to
whether it wants food or water, and they are confident that most
people, rather than claiming that pigeons use language, will be
content to admit that many of Lana's stock phrases owe nothing to
linguistics and everything to simple rote learning. 'The rich
linguistic knowledge implied when a child utters a four-word
sentence such as "Mommy give me milk" is apparent when we
consider other sentences the child is able to generate,' Terrace
explains. ' "Sally give me soda", "Daddy give cat milk", "I throw
dog ball", and so on. In any of the four positions of the original
sentence, the child can substitute a variety of words, each
appropriate to the circumstances at hand. Such sophistication was
not demonstrated by chimpanzees,' he concludes.[9]

Indeed a major part of the problem of teaching apes language is in
the name, or gloss, that we give to a sign or symbol, as the example
of Terrace's pigeons shows. When Francine Patterson's signing
gorilla Koko bites Patterson and then signs 'Koko bad Koko sorry',

are we justified in thinking that she has any real understanding of the concepts that we mean by 'bad' and 'sorry'? And when Nim signs 'time', is it because he has some concept of time, or because he has learned that 'time eat' is an effective string of signs, effective in that it might get him what he wants? The word 'cry' is a particularly ironic instance. Many of the signing apes use the word in a completely understandable way, to indicate things that might be painful or cause anguish. But apes do not cry; only man has the capacity to weep, so the apes can have no appreciation of what it means to cry, though of course they can understand and feel pain.

In the final analysis it seems that the chimp language studies have shown not that chimps can acquire the rudiments of language but that they can acquire a sophisticated number of tricks and talents that they perform more or less appropriately in order to secure rewards rather than to communicate for its own sake. Of course, this is no mean feat; much of what passes for communication between humans is no more. Even so, the initial uncritical enthusiasm for, and acceptance of, the ape language studies has now given way in many quarters to an equally unthinking dismissal of the apes' abilities. The backlash against the ape language studies has at times been harsh and acrimonious, but out of it has come a new awareness of the care that must be taken in many experiments before we can be sure that our interpretation is correct. And from those new experiments has come evidence of rather startling abilities possessed by apes. The philosophy behind some of the studies has also been examined, with beneficial consequences. All along it seems that the advocates of apes' linguistic strength have been eager to dethrone man from his self-awarded seat, at least as far as that throne is represented by language. And the critics have been just as eager to keep man high on his linguistic pinnacle. Both sides have paid insufficient attention to the real skills of the apes: *we* are proficient at language; what can *they* do?

The truth is that apes are very good indeed at communicating, but they don't use language to do so. We have purposely avoided giving a definition of language ourselves, not least because we feel that the to-ing and fro-ing over definition is one of the more sordid aspects of the controversy over ape language. But it must be clear that in the centre of the fuzzy area that is language we are all aware of a hard kernel and are all aware of what we mean. The dance language of the bees, for example, whereby eager scouts who have discovered a rich new source of food encourage their sisters to exploit it and tell them exactly where to go, is neither a dance nor a language. We call it a dance language because that is a handy way to refer to it, a way that, because of the properties of our language, instantly communicates some of the vital features of the bees' system; it involves repetitive rhythmic movements and conveys information. What then *is* the dance language? It is a remarkably sophisticated communication system that certainly subserves one of the functions of language, namely the transfer of important information from one individual to another, but there is more to language that this. So too there is more to the communicative abilities of chimps than we may have suggested in the preceding pages. Aside from the mountains of data collected by Jane Goodall and the scientists who followed the trail she blazed among the chimpanzees of Gombe Stream in Tanzania, and the corroborative insights gained by other field primatologists at other sites, there is a powerful amount of evidence from a group of captive chimpanzees watched by Emil Menzel in the wilds of Long Island, New York.

Menzel is a research psychologist whose approach to understanding the chimp is in the tradition of the classical ethologists. A tall and gentle man, he explains how his aim is to 'look at the society as a whole and try to discover where it is headed and what are its goals. Then you can say what the individuals might have to communicate about, and ask how they do it'.[10] Menzel has a large, relatively luxurious paddock for his chimps, who were born in the wild but brought together while still young so that they form a cohesive social group not noticeably different from groups of wild chimpanzees. The 500-by-100-foot field contains all sorts of hiding places and interesting objects for the chimps to use but is bounded with a high fence that is topped by an electrified wire. The fence is designed to keep the chimps away from the outside world, which contains even more interesting things for them to use, but it sometimes fails singularly to do so.

Menzel spends a great deal of time just watching his chimps, recording what they do and attempting to make sense of it. But he also does experiments — nothing elaborate, just a little perturbation to the system to see how it works — and the most fascinating of these are his investigations of communication between the chimps. His method is simple, but ripe with possibilities. First he hides a number of objects — food, or toys, or even rubber snakes — at various locations around the paddock. Then he locks the chimps into a small cage at the edge of the paddock, from which they cannot see out into the field. The experimenter then takes one of the chimps from the group and, away from his fellows, carries the chimp around the paddock showing him the various hidden objects. The chimp, who can fairly be called the leader (though he may well not be the dominant animal), is put back with his chums in the cage, and after a few minutes together the group is released in the paddock to do as it pleases. What pleased it most was to find hidden supplies, and Menzel and his team watched closely to see if they could detect how the chimps communicated with one another.

The results were very striking. In the simplest case, when the leader had simply been taken round several piles of food, the group set off round the paddock and quickly discovered all the food. They never searched locations that had not been baited, and they never missed a baited hiding place. They didn't return to a hiding place that they had already uncovered, and unless the demonstration route had been a very efficient one they followed their own route between the hiding places rather than simply mimicking the path taken by the experimenter when he had been carrying the leader around. The leader had, at the very least, a clear psychological map of the paddock with the location and nature of the hidden objects marked upon it, and his fellows also seemed to have ready access to this information.

The leader did not use any grand gestures to get his comrades to follow him. It was all done with subtle signals; a glance here, a shrug there, a nod of the head or a tap on the shoulder. As Menzel notes, 'the most dramatic and humanoid looking signals were made by the most infantile and inefficient leaders, and they decreased markedly as the animals gained experience at leading'.[11] The leader could also communicate the nature of the hidden objects; if it was food the chimps advanced straight to the spot and felt around in the grass with their hands, but if it was a rubber snake in the grass they came forward gingerly and kept their distance, fanning out and sometimes searching out a branch with which to poke among the long greenery. Often the leader was not in the van of the group but was in the middle or even at the back. His companions had divined the information and didn't actually need the leader himself to lead them to the spot, only to tell them where the spot was. Two leaders, shown different-sized piles of food, attracted followers in proportion to the pile of food they were going to, and leaders going to preferred food (fruit) attracted more followers than leaders going, at the same time, to vegetables. All in all, Menzel's chimps showed a remarkable ability to spread information around the group.

Perhaps this isn't so surprising. Many of Jane Goodall's stories from the wild chimps of Gombe are very similar, in that chimps with information are able to impart that information to their companions with a subtlety and accuracy that makes any definition of language irrelevant.[12] The Gombe chimps were similar to Menzel's in other ways too; both groups could lie. One day Figan, a young Gombe male, spotted a banana that Goodall had placed in a tree. Below the tree sat Goliath, a high-ranking adult male. Goliath would surely exercise his right to the banana if he were to become aware of its existence, which he would if Figan either approached or stared at the banana. Figan must have worked all this out, for he deliberately abandoned the tempting tree and loitered beside one of Goodall's tents, idling away a quarter of an hour until Goliath left and the young schemer could have the banana to himself. No wonder Figan eventually became top chimp.

Menzel's chimps would do the same sort of thing if the pile of food they had been shown was a very small one. The leader would try to supress the information until he was on his own, or he might lead the others off on a false trail and circle back when they weren't paying attention. But followers are also very adept at seeing through a leader's feigned indifference, and the less eager and obvious he tries to act, the more closely the followers keep him under surveillance. And Premack's chimps could also learn to lie. In one experiment they watched as food was put into one of two boxes. A trainer – the 'good guy' – entered, and if they could somehow indicate to him which was the correct box, he would open it and share the food. If the trainer got it wrong, neither of them got the food. A different trainer, the 'bad guy', would also open the box, but he wouldn't share the loot. After some effort – the chimps seemed to be guileless – the animals learned to point at the wrong box, thereby getting all the food for themselves. By their actions they had to lie about the location of the food, and they succeeded in doing so.

Young Gombe males often elicited support from one another to tackle a project too big for any one of them alone. This might be an attack on a neighbouring group, or perhaps a hunt for a young colobus monkey or a bushbuck. In a similar way, Menzel's chimps banded together to break out of their paddock in New York. One ape learned to pole vault out of the enclosure; the tradition spread like wildfire through the simian culture and soon they were all vaulting the fence. When that escape route was made more difficult they learned to prop poles against the trees that overhung the compound and thus to climb to a temporary freedom. Indeed, in one photographed case a chimpanzee solicited the help of a companion to lift such a pole onto the fence. Then the instigator sat holding the foot of the pole secure while his helper escaped, and obligingly stayed there while a third companion, who had not been part of the escape team, also climbed out. Only then did the instigator climb his ladder.

So it is clear that chimpanzees can communicate intentionally with one another, and can gather information even in the absence of specific communicative signals. If they could not, they would not be able to conduct their complex social lives as they do. To say that their communication is not language doesn't really tell us anything other than the fact that they are not human beings. Our communication system – language – developed in response to evolutionary pressures that made the ability to transmit information about certain things an advantage. It is very hard to know just what those pressures were; at one time, in keeping with ideas about the hunting origins of man, it was said that language was needed to organise and co-ordinate the hunt, but this cannot be, because human hunters don't talk while hunting and because other animals, such as wolves, lions and hunting dogs, seem to be able to organise a co-operative hunting and social life without benefit of speech. Perhaps the development of language was just one of those inevitable things, a by-product of the expansion of the brain, but this is not a very satisfactory answer at all.

Most probably language helped the early ancestors of man to run their particular advanced food-sharing society. If men were doing one job and women another, they would need to tell one another what they were up to. As Glynn Isaac explains, 'the adaptive value of food-gathering and division of labour would be greatly enhanced by improvements in communication; specifically, the passage of information other than that relating to the emotions, becomes highly adaptive.'[13] Isaac buttresses his claim by referring to the other food-sharing societies, such as the honey bees, that have developed systems that are quite close to language, and the case that he makes is a persuasive one.

A different primate, the vervet monkey, lives in more open savannah than the chimpanzee and while it doesn't share food it does have what might be described as a proto-language. That is the conclusion of Dorothy Cheney and Robert Seyfarth, two likeable young research workers from Rockefeller University in New York. They had read that vervets use a different sounding alarm for different predators: a leopard gets a series of staccato barks, a bird of prey a sort of 'rrraup', and a snake a chutter. What is more, the responses to the alarms are eminently sensible. If the monkeys are on the ground when a leopard is spotted, they run into the trees, a good idea as leopards lurk in the bushes. Seeing an eagle, the monkeys rush out of the trees and onto the ground, from which the eagle cannot take them. And a snake causes them to stand on tiptoes and peer all around; if they spot the snake they all rush over and mob it, harassing it until it slinks away. But are the alarm calls in any sense words that refer to the predators rather than the alarmist's emotions? What Cheney and Seyfarth did was to make tape recordings of the alarm calls and play them back to the monkeys when there were no predators in sight. The monkeys responded to the false alarms as if there had been a real predator, rushing into the trees for a leopard call, coming out of them for an eagle call, and peering about for a snake call. The different alarm calls did seem to be proto-words, with a meaning. 'The important thing,' Seyfarth says, 'is not the exact meaning of each call but the fact that the monkeys are clearly using different vocalisations in a way that effectively represents different objects of events in the external world. This raises a multitude of questions about the way we study animal communica-tion, and indeed about the evolution of human language itself.'

'It's obvious to anyone,' continues Cheney, 'that human language is a unique form of communication, and we generally assume that when hominids evolved the first language they gained a considerable selective advantage over other species. One of the things that language allows us to do is to signal to each other about different objects or events in the world around us. Now, if we assume that vervet monkeys have developed such an ability − admittedly in an extremely rudimentary form − and if semantic signalling confers such a great selective advantage, why wouldn't the same monkeys also signal about objects during their more relaxed day-to-day social interactions with one another?'[14] It's an intriguing question, one that Cheney and Seyfarth returned to Amboseli National Park in Kenya to investigate. The data are in, and while it is too soon to be certain, it does appear that some of the softer grunts that the monkeys employ in everyday situations (and which are indistinguishable to the human ear) are indeed being used as protowords.

If vervet monkeys actually use, in the wild, something akin to language, does that mean they are more human than chimps? Of course not, not least because nobody has studied chimps with the intensity that Cheney and Seyfarth devoted to the vervets, but also because there is more to being human than having language. You wouldn't think it, though, from the emphasis that people put on speech. An interesting point to ponder is why the English word 'dumb' means both 'unable to talk' and 'stupid'. Another interesting point is why the extravagant claims of the chimp language studies found such ready acceptance. David Premack put it this way: 'Chimps do not have any significant degree of human language and when, in two to five years, this fact becomes properly disseminated, it will be of interest to ask, Why were we so easily duped by the claim that they do?'[15]

We also like to picture ourselves as rational, thinking beings. Some of the work to have emerged from studies of ape language has revealed that chimpanzees too are capable of quite sophisticated reasoning and logic. The best of this has come from Premack's group at Philadelphia, which has used its experience of training chimps to employ token words to devise tests of reasoning. At base, perhaps, the question is whether language is necessary for reasoning. This is an old philosophical debate, one that is probably not resolvable, about the nature of language and the nature of thought. Do we think in words? Or concepts without names? It seems obvious to non-philosophers that both are true, but the chimpanzees provide a way to test some of the philosophers' propositions. Accepting, for the moment, that although chimps can communicate they do not have human language, one can discover whether they can perform the sorts of reasoning task that are often said to require language in man.

Douglas Gillan, a quiet, slow-talking Nebraskan whose manner belies his perspicacity, did a great deal of work with Premack's Sarah and the other Philadelphia chimps to get them to complete tests of reasoning. One of the first things he tackled was the notion of logical inference. At its simplest, this means the ability to perceive the relationship between things. For example, if we tell you that a furlong is longer than a chain, and a chain is longer than a pole, you should be able to work out that a furlong is longer than a pole. We didn't tell you that the furlong was longer than the pole, you inferred it. To get at this problem in chimps, Gillan set up an ordered series of coloured boxes. Over a number of trials, the chimp was offered a choice between two boxes, one of which contained more food than the other. The colour that contained the food was specified by a series that Gillan had drawn up. So, for example, red might be higher than green but lower than blue; when red and green are together the red box is the one to choose, but when red and blue are together it is the blue box that is 'higher' and contains the food. Patiently Gillan put the apes through their paces, offering them choices between adjacent colours in the series. Apart from the top and bottom colours, each colour was positive, that is contained food, on the same number of trials as it was negative, that is was empty. Finally came the big day when the chimps were offered choices between nonadjacent colours, blue against green, for example. 'They got ten out of twelve right,' Gillan told us, 'and don't forget that they'd never seen the whole series. We didn't ever let them see nonadjacent members of the series, they had to figure it out for themselves. They had to infer the logical structure of the series. And they did.'[16] Sarah, who had some language training, was one of the successful chimps, but others, who had not been through the same procedures, did every bit as well. Chimpanzees can make transitive inferences, and they can do so without benefit of words.

Gillan looked at other forms of logical reasoning too, asking whether chimps could reason by analogy. This sort of test is rather like the kind of thing that is common on so-called intelligence tests: glove is to hand as shoe is to . . . ? It was tests like this that led to the conclusion that Sarah could master cause and effect, for she correctly perceived the analogy between knife and apple slices, water and wet sponge, as examples of cause and effect. Gillan is enthusiastic about the work and what it reveals: 'I think it tells us something first about reasoning, and that is that reasoning does not require language. Language is not a necessary condition for at least some types of reasoning . . . [which] can occur in species that don't have language.'[17] And from that basis the Premack group has gone on to ask all sorts of interesting questions about the nature of the chimpanzee mind.

'That sounds a little obscure and even a little weird,' Gillan told us, 'so let me give you an example of something. One kind of task that was given to Sarah involved showing her films of a human, one of her trainers, facing a difficult problem. For example, he might be locked up in a cage and trying to get out. Sarah saw this film of a person locked up in a cage and trying to get out, and then she was presented with various solutions, for example, a picture of a key and a picture of a lighted match. She consistently chose the object that would get the person out of the problem, in this case the key. One of the interesting things is that she seems to have seen that this videotape scene that she watched was more than just a videotape scene but that it had elements of a problem. Another interesting thing is that in order to choose the right object, the key, Sarah had to put herself in the actor's place. She had to see things from his point of view, and had to have, in a sense, a theory of that person's mind. That's a complicated idea; she would have to make certain assumptions about what his state of knowledge was, she would have to make certain assumptions about what he could see, and couldn't see, and there are a number of interesting implications for this sort of idea.'[18]

There certainly are, for if a chimp can, under the right circumstances, demonstrate such powers of thought and reasoning, then it is far more than the dumb brute we like to imagine. The Rumbaughs too have gone beyond their original rather poorly conceived studies of how the chimps used Yerkish to speak to the computer, which came down in the end to a sophisticated set of rote phrases, learned parrot (or pigeon) fashion. Like Premack, they have adopted an atomistic approach, splitting language into atoms of structure that they feel can be examined in detail. One atom that the Rumbaughs have looked at in this way is so-called reference, which they regard as 'the linguistic essential'.[19] The three stars of the Yerkish project, Lana, Sherman and Austin, were taught two new pictograms. One, translated as 'foods', applied to oranges, bread and beancake. The other, 'tools', was for keys, money and sticks. Once that part had been mastered, the Rumbaughs asked whether the label really did mean 'food', or whether it simply had been attached to the three specific foods. They tested the chimps with five new foods, and five new tools. Poor old Lana was no good at this, getting only three out of ten. But Sherman got nine out of ten (he incorrectly labelled sponge — 'the tool which he occasionally eats portions of as he uses it'[20] — as food) and Austin scored 100 per cent, labelling all ten new foods and tools correctly. Lana dropped out at this stage, but Sherman and Austin went on to attach the label not to the actual objects but to photographs of the objects. This learned, they finally moved into the last stage of this project. They learned to associate the symbolic pictogram for food or tool with the pictograms for the training foods and tools. Now they were tested to see how they would label the pictograms of test foods and tools. Austin again scored 100 per cent, and Sherman once again blew it by labelling the symbol for 'sponge' as a food. We are tempted to say that, no matter what Sue Savage-Rumbaugh may think, as far as Sherman is concerned the sponge *is* a food, not a tool, but our objection is neither here nor there. The point is that in learning to label lexigrams correctly, the chimps had shown a high degree of intellect. To succeed, they had to go through a chain of thought that began with recognising the lexigram, then required that the object associated with that lexigram be brought to mind. That object had to be characterised as a food or a tool according to functional characteristics learned with other objects of the same class, and finally the correct lexigram for that class had to be remembered and selected. Again, quite an achievement, even for a chimpanzee and even if one does not believe that chimps use language as we do.

Interestingly, language, while it has always been the main barrier between ourselves and the animals, has taken the place of another ability — tool-use. Dr Johnson defined man as the tool-making animal, and some palaeontologists reserve the genus *Homo* for animals that they believe made tools. People feel that somehow the manufacture and utilisation of tools demands a form of intellect qualitatively different from anything else. Now that we have the records of observers who have spent many patient hours watching animals in the wild, tool-use has faded somewhat from the forefront of exclusive characteristics. A whole host of animals, including even snails, has been seen using tools, and quite a few species modify objects in the environment and so can be said to manufacture tools. Man the tool-maker is no more, despite the assertion of anthropologist John Buettner-Janush (subsequently jailed for manufacturing not tools but illicit drugs in his laboratory) that 'man makes and uses tools in a distinctly different way than chimpanzees.'[21]

What of ape the tool-maker?

Early in 1971, less than a decade after Jane Goodall's revelations of

tool-use in the chimps, R. V. S. Wright contemplated an iconoclastic experiment. What if, instead of termite fishing, Goodall had
seen her chimps pick up a stone, chip flakes from it, and use the chips
to cut through, say, meat? He set to, sitting in a cage with Bristol
Zoo's orang-utan Abang. Wright showed Abang a food box tied
with nylon rope, and showed the ape how to cut through the rope
with a flint. Even hampered by a short thumb (an adaptation to
brachiation) Abang quickly mastered the precision grip and learned
to use the sharp flake. That done, Wright moved on to making the
tool in Abang's presence. He struck the flint core repeatedly, flakes
flying. Abang apparently enjoyed the noise, but it took quite a while
for the ape to grasp what was required. Eventually, by the tenth
session, Abang would go through the whole sequence. Presented
with a tied food box, flint, and hammerstone, Abang would knock a
flake from the flint and use it to saw through the rope.[22] Now of
course Abang had not invented tool-use but had only imitated
Wright's abilities. Even so, the fact that such a task could be
mastered, and by an ape more suited for arboreal brachiation than
terrestrial walking, is testimony to the intelligence and intellectual
powers of apes.

The picture emerging from all the studies of apes' abilities is that
these animals possess many of the abilities that naive people regard as
uniquely human. They may not speak, but they can communicate
very effectively. They may not write philosophical treatises, but
they can reason logically and probably are able to do far more in that
line than man will ever be able to discover. Their social lives, which
we have not dealt with in this chapter, are complex and
Machiavellian, constructed on different principles to ours but similar
in being dependent on a web of relationships. What are we to make
of all this? The simplest attitude to adopt is to assume that the
ancestor of man and chimpanzee was very like a chimp, and that since
they split man has changed a great deal and the chimpanzee not at all.
A problem with this is to explain why our ancestor should have been
like a chimp rather than a gorilla, and why man and the gorilla
should have changed but not the chimpanzee. There are no grounds
for preferring the chimpanzee, rather than the gorilla, as man's
ancestor, except perhaps that because it occupies more accessible
country we know more about the chimpanzee, and so our
speculations are more firmly based. Perhaps the only reason not to
propose the gorilla as the ancestor is that it is more specialised, living

a secluded life and eating a prescribed diet that somehow does not seem to be how we imagine early man to have been.

This view cannot be correct. Anatomically the ancestor of the African apes and ourselves that made the trek out of Asia may well have been more like a chimpanzee than like a man, but we have shown that it is entirely possible that the direct ancestor of the three modern species — *Homo*, *Gorilla* and *Pan* — was in many respects quite man-like. And it is a startling fact that, despite the wealth of information gathered from the wild, the potential that the apes reveal under man's tutelage far exceeds anything seen in the wild. Nim, Washoe, Austin, Sherman, Sarah, Sadie and all the other performing apes astound precisely because their abilities are unexpected. Nothing we know of their life in the wild prepares us for the skills they display. They seem to possess prehuman abilities that are of no use to them in the wild and so are not used.

This seems like a very foolish thing to say. How do we know what the apes do and do not use in the wild? We admit that we cannot be sure that there are not chimpanzees passing their time absorbed in analogical reasoning or transitive inferences. Many of the skills that amaze us are probably vital for the smooth running of society. As Doug Gillan says, 'It would save you a lot of time and bruises if you could figure out that "because B beat me up and A could beat B up, I'm not going to mess with A".'[23] That, perhaps, provides a clue to the evolution of those skills in man. Perhaps what we see in the chimpanzee and gorilla are the rudiments of intelligence as developed by our ancestral brachiator. What they have is a seed that was not nurtured, because the lifestyle that they pursued did not require the fruits of that seed. Ape society is simple compared to man's, but it is complex enough. In such a society, the ability to make inferences, the techniques of communication, and the need to be something of a psychologist, to understand other individuals' points of view, would be extremely useful assets. They are precisely the features we find in apes, and they are the features that man has built upon. Just as we infer that anatomical similarities imply an ancestor with those features, perhaps the cognitive similarities of man, chimp and gorilla imply an ancestor with those abilities. Then, after the three species had gone their separate ways, one species, our own, was selected for extensions of those cognitive abilities, while the others found no advantage in being able to talk. An organ without a use, for example the appendix, does not quickly vanish. It must be positively costly for selection to remove it, though of course it may atrophy, as man's appendix has done. Similarly the cognitive abilities of chimps and gorillas may represent a sort of appendix of the brain, although they are far more useful to the apes than our appendix is to us.

Chimp society is less complex than our own, and chimp psychology is less complex than our own. Which is cause and which effect we do not know. The safest bet is that, like so many other facets of hominid evolution, the two are inextricably interwoven. Complicated society requires a better brain, and a better brain enables a more complex society. Complex society enables one to cope with a particular sort of changeable environment, and the environment requires that society if the animals are to survive. On their path, the chimpanzees and gorillas did not exploit the niche we did, and naturally they did not adapt to our niche. The skills they have are the skills that they need, and the skills that we have are the skills that we need, and no amount of cramming or teaching can turn one of the African apes into the other. That may sound like a truism, but it is none the less true for that. The history of man's efforts to understand the apes and through them himself is littered with examples of people totally failing to remember that we are different species, evolved to fill different niches, with our own abilities and specialities.

We have concentrated on efforts to teach apes human skills. Turning the tables for a moment, let us consider one ape skill, that of termiting. Thanks to the painstaking efforts of Jane Goodall, it is now practically common knowledge that chimpanzees use tools, and one of the tools they use is a thin twig with which they 'fish' for termites. The termites only emerge in two months, October and November, so the chimp first has to know that it is open season and that fishing will be rewarded. If termiting will be successful, the chimp approaches a termite mound and selects a tool, a fishing rod. This is usually a piece of vine or a small twig, and it may modify the tool by stripping off some of the leaves. Good fishing rods may be brought to a termite mound from some distance away, and may be brought in batches of three or four, in anticipation of a good day. The next stage is to find, on the surface of the mound, one of the sealed entrances to the termite colony. The chimp scratches the entrance to the tunnel away, and inserts the fishing rod. After a little twiddling, the chimp withdraws the rod, covered now with termites, and strips the termites from the rod with his lips. The procedure will be repeated time after time for hours on end; occasionally repairing a damaged rod or selecting a new one, the chimps almost effortlessly gain valuable supplementary protein. But while termite fishing may have removed another of man's picket fences, that of the tool-using animal, it didn't seem all that difficult. Geza Teleki, one of the scientists who followed Goodall into Gombe, decided to try for himself.

The first thing Teleki discovered was that he couldn't even find the sealed tunnel entrances. The chimps unerringly scratched them open, but he could not. That alone revealed 'knowledge far beyond my expectations' Teleki conceded, and he suspected that they must memorise the position of the hundreds of hidden openings on each mound. Next came the business of making a fishing rod. The requirements seemed simple; the probe must be supple enough to bend with the turns of the termite tunnel. A chimp did not seem to select any probes that were not suitable, but Teleki often found that a twig or branch that he thought would be perfectly good was not. The final stage, inserting the rod and withdrawing it loaded with tasty termite soldiers, would be easy, or so Teleki thought. For weeks he enjoyed 'nearly total failure', until he finally 'began to appreciate the problems involved'. Chimps insert the probe with a deft subtlety and vibrate it gently. This baits the soldier termites to bite the probe. Then the chimp withdraws the rod smoothly and gracefully, so as not to separate the termites from their jaws. Teleki was an almost total failure, judging himself to be about as competent as a four-year-old juvenile chimp. That is, not very. 'Incompetent as they were,' he reports, 'my attempts to acquire the skills of locating tunnels, selecting materials, inserting probes, and extracting termites left me with a healthy measure of respect for chimpanzee technical ability.'[24]

The point we are trying to make is a simple one. Nobody knows what the common ancestor of man, chimp and gorilla was like. We do not know what it looked like, how it behaved, whether it could do some of the things that modern apes can. Looking to the chimpanzee or gorilla for direct guidance to our past is therefore doomed to failure, because both species have evolved away from the common ancestor as surely as we have. They may not have moved as far in some directions as man, so that their abilities in certain spheres are less adequate than ours, but they have adapted to their own niches where their skills are more suitable than ours. Man cannot fish for termites, but unlike chimpanzees he has learned to beat the mounds with leafy branches, simulating the sound of rain and causing the termites within to swarm out and be caught. That is evidence of man's flexibility and cleverness, his own speciality.

We have taken a slight detour on the road to man to examine the species that took different paths. We have not looked at them in any great detail, but we hope nevertheless to have shown that they are close to us in many respects other than biochemistry. We have hardly discussed the question of why man and the African apes pursued such different paths after they split, because in this book we are more concerned with establishing the fact that the split was recent than with any particular set of events after the split. The three species of African ape — man, chimp and gorilla — differ hardly at all biochemically, a little more anatomically, and a great deal psychologically. Despite the manifest similarities, the differences are what concern most of us. Those differences arose in a matter of a few million years. Perhaps one day we shall learn how and why.

Ten
THE MONKEY PUZZLE SOLVED

The case we have made seems so secure that we hesitate to press it further, for fear that it may seem that we protest too much. And yet, many people find the conclusions disturbing, and even within scientific circles the new picture of mankind's evolution provided by the molecular evidence has not received the acclaim it deserves. Perhaps, after all, it would be wise to offer a final summary of our interpretation of the molecular evidence, and to address also the question of just why this startling revision of the timescale of human evolution has lain practically dormant for fifteen years. Our starting point is the fact — not hypothesis, or even theory, but a measured piece of evidence — that man, chimp and gorilla are not only very closely related, but are all equally close to one another. There is no way you can group any two of them together to the exclusion of the third on evolutionary grounds.

Vincent Sarich, now a joint professor in the departments of anthropology and biochemistry at Berkeley, has an awkward question to put to anyone who still suggests that the human species did not evolve from an ancestral form. 'If we were specially created,' he asks, 'then why were our genes created in the image of those of the African apes?' Thanks in large part to Sarich's ground-breaking work, we now know that human genetic material is 99 per cent the same as the genetic material of chimpanzees and gorillas. And the reason that the three of us are so similar is because until very recently our genes did not exist as separate species but dwelt together in a shared common ancestor. The three lineages only became distinct, to follow their separate evolutionary paths, less than 4½ million years ago. This fact is possibly the single most significant discovery pertaining to the mystery of human origins and evolution since Darwin's own day.

There is no longer any room to doubt that man evolved from an ancestor that he shared with the gorilla and the chimpanzee. The evolutionary explanation is established. But there is plenty of room for doubt and debate over the traditional, and now establishment view of what exactly the human species evolved from, and when, and where. We believe that, as well as providing the proof that we share a recent common ancestor with the chimp and gorilla, the evidence of the molecules provides the key to these remaining mysteries.

Humans and chimps (and gorillas) are about as closely related as dogs are to foxes, horses to zebras, walruses to sea-lions. They are as close to each other as gophers on one side of the Colorado river are to members of *the same species* on the other side. This very close relationship — some taxonomists would describe the three African apes as sibling species — simply cannot be explained within the framework of the standard picture of human evolution. The establishment's pictures have received a lot of attention lately, but without wishing to denigrate the masterful efforts of the palaeontologists, the details of those pictures owe more to imagination than evidence. The fossil evidence of man's ancestry is much more sparse and incomplete than many people realise. Christine Janis, a palaeontologist who specialises in understanding the development of the jaw and chewing apparatus of herbivores, has worked in many of the major palaeontological centres, including Cambridge Massachusetts and Cambridge England. She points out that 'there are more people working on fossil primates than there are fossil primates. And to put that in perspective,' she adds, 'there are more than ten thousand fossil horses in the Frick collection of the American Museum of Natural History and not more than a handful of people working on them.'[1] Nobody has counted the fossilised primate fragments, but clearly she is referring to substantial remains, and there can't be more than a few dozen of those.

So the fossil evidence of human origins is sparse. Is it reliable? This is a touchy point, one that palaeontologists are sure to deny, but it is an uncanny fact — amply documented — that fossil-hunters have had a happy knack of finding exactly what they were looking for.[2] Even more disturbing, the traditional picture of human evolution is not based on the same sort of analysis as would be applied to any other group. Much depends on statements from on high, from the few acclaimed gurus of palaeontology. With egocentric interest in ourselves common to us all, those statements are elevated in importance and take the place of hard evidence. Allan Wilson, Sarich's mentor and partner, is clear about the nature of palaeontology. 'There's a lot of climate of opinion that so influences you, even somebody like me who's tried to be outside the anthropological world and as critical of it as possible. I was taken in by the climate of opinion, which said that "we know the branching order for these apes and humans" and [I thought] that somebody must have really gone through carefully and constructed a tree based on a comprehensive list of traits. And nobody's ever done it.'[3]

Wilson's accusation strikes at the heart of the matter. The traditional view of mankind's origins, and the placement of the branches on the evolutionary tree, *has never been properly constructed or justified*, as it would have been if the fossils were pigs rather than primates. The morphologists have not properly addressed the question of man's relationship to the other primates. 'We're the first people who addressed that question and have produced a set of data,' Wilson contends, 'and we now have a set of data that's much larger than I think has ever existed. The morphologists can't come close to us.'[4]

The molecular clock, remember, provides us only with a ratio of the times that have elapsed since various species shared a common ancestor. It tells us, for example, that human and baboon last shared a common ancestor 4.5 times as long ago as human and chimp, and just over half as long ago as human and squirrel monkey. The puzzle now is whether the human–chimp split was 2 million years ago, say, which would put the human–baboon split at 9 million years ago, or whether the correct figures are nearer 20 and 90 million years ago. To decide, we have to turn to the fossil record. The molecules don't provide times directly, so, as Sarich says, 'you set it [the clock] with reference to the fossil record. There isn't any other way.'[5] The new timescale of mankind depends as much on fossil evidence as on the molecules; the two are inextricably entwined, and none of the fossil evidence contradicts the new timescale. Of course, if the fossil record were perfect we would have accurate dates for all the various splits, and the molecular evidence would merely be corroborative, as it is in so many other animal groups. But for the primates the fossil record is imperfect, and the molecules play a key role. Years of study have confirmed that the clock ticks at the same rate in all primate lineages, so that one good fossil date is all that is needed to calibrate the clock — in fact we have several good dates and they all fit into the same framework.

The results that come out of this calibration are so surprising to the traditionalists that they have met with stiff opposition. In the face of that opposition the molecular biologists, unused, perhaps, to the vigour of debate in palaeontology, have not always pressed their case as strongly as they might. 'Neither of us has really gone on the campaign trail,' Wilson admits. 'The tradition in biochemistry, at least at Berkeley, is that the ideas and the facts should speak for themselves.'[6] In trying to please as many palaeontologists as possible — or rather, as Sarich puts it, to make 'the largest number of paleontologists least unhappy'[7] — they may have made their data look less clearcut than they really are.

For example, most traditionalists would be extremely unhappy with any suggestion that man, chimp and gorilla shared a common ancestor less than 10 million years ago. So they take this date and use it to poke fun at the molecular clock by showing that, if man and the other African apes did split 10 million years ago, then, according to the molecules, the split between rat and mouse occurred 75 million years ago. 'Everyone knows,' they proclaim, that the rat and mouse actually split just 5 million years ago, and that proves that the molecular clock is useless.

Not quite. First of all, there is no good evidence that the rat and mouse did actually split only 5 million years ago. That seemed a good bet at the time this argument was first aired, but it stemmed from a guess by a single noted palaeontologist, working without the benefit of any good fossil evidence. Even so, let's follow the argument through and see where it gets us. Timing the man–ape split at 10 million years ago (twice as long ago as the actual date revealed by the molecular clock) gives an incorrect date for the split between rat and mouse. But using the accepted palaeontological date for the man–ape split, 20 million years ago, gives a ludicrous date for the rat–mouse split of 150 million years ago, back early in the age of the dinosaurs, before mammals, let alone rodents, had appeared on the face of the Earth.

That, however, certainly does not prove that the molecular clock is inaccurate. If the palaeontologists wish to follow this line of argument through logically, they must also take their preferred date for the rat–mouse split at face value and examine the implications of that for the timings of the clock. When we put the palaeontologists' rat–mouse date in as the calibration, we find that the date for the split between man and the African apes leaps forward to only 600,000 years ago. Clearly, something has to give — but the inconsistency might be resolved if new evidence came in to change *either* of the key dates, the man–ape split or the rat–mouse split. In the past few years, since this argument first reared its head, new fossil evidence has come to light and has led experts to revise their date for the rat-mouse split back from 5 to 15 million years ago; to fit with the 4½ million year date for the man-ape split (and taking into account some known peculiarities of rodent evolution), the rat-mouse split ought to be somewhere over 20 million years ago, and palaeontologists are still pushing the figures back. 'It's not going to get less,' says Sarich, 'and I think 20 or 25 is not in the least an unreasonable figure.'[8]

So even the rat-mouse split does now seem to fit with both the fossil evidence and the timings of the molecular clock. The traditionalists who held the 'patently false' figures that the molecules gave for the rat-mouse split up for ridicule have found their arguments rebounding on themselves, for the figures are no longer patently false and, indeed, look eminently sensible. The implication is that the figure for split between man, chimp and gorilla is also eminently sensible. Taking many well established fossil timings into consideration, the molecular clock can be read to fill in the blanks in the record. The combined result is a firm series of dates.

The placental mammals originated and began to diverge from the marsupials no more than 100 million years ago, which sets an upper limit on all the rest of the important events on the human family tree. The origin of the primates cannot have occurred more than 75 million years ago; the primates of the New and Old Worlds diverged no more than 35 million years ago (comfortably agreeing, by the way, with geophysical evidence for the break-up of the former supercontinent); Old World monkeys split from the hominoids no more than 20 million years ago, and the modern ape family began to radiate approximately 15 million years ago; gibbon and orang-utan diverged 11 and 8 million years ago and finally (ignoring the more recent divergences into pygmy and modern chimp, mountain and lowland gorilla) there was the three-way split between man, chimp and gorilla no more than 4½ million years ago.

These dates are especially interesting, as we have pointed out, in view of the environmental changes that were occurring around the world at that time. After many millions of years of relatively warm and equable weather on Earth, between 4 and 2 million years ago the steady process of continental drift was bringing the land masses of the northern hemisphere into essentially their present positions; the continents surrounded the North Pole, restricting the flow of warm equatorial ocean currents and allowing an Ice Epoch to begin. By an incalculably rare coincidence, the South Pole is also covered with ice, the outcome of a continent having drifted onto the pole. As a result the Earth, for the past 2 million years, has been in the grip of a full Ice Epoch, broken only by short interglacials, like the one we are living in now. These interglacials come and go with the astronomical rhythms of the Earth's orbit around the Sun, and they last between 10 and 15 thousand years. The interglacial we inhabit has already existed 11,000 years.

The changes in climate did not bring ice to the valleys of East Africa; but they did bring drought, which set in at just the time that the split between man, chimp and gorilla was taking place. Climatic and environmental changes are always associated with evolutionary upheaval, and the present epoch is no exception. To geologists 100 million years hence (if there are any) the times we live in will be regarded as a geological boundary every bit as significant as the end of the Cretaceous, 65 million years ago, when the dinosaurs (and many other species) disappeared from the face of the Earth. The boundary between the Tertiary and Quaternary periods is set at the beginning of the most recent Ice Epoch, about 3 million years ago. Three million years seems an awesomely long time to a creature that measures its life in decades, but it is a fleeting moment in geological time, and we are far too close to the boundary to see it clearly. Like the end of the Cretaceous, when the demise of the dinosaurs opened the way for the small mammals that had survived the disaster (whatever it might have been) to radiate out into new ecological niches, creating new species and groups from the diversification and branching of the old stock, so the great Ice Epoch has opened up new opportunities for those species able to take advantage of them.

By 4 million years ago our ancestors were already very man-like and well on the road to human intelligence and a co-operative, sharing society. The discoveries made by the Leakey family in East Africa, as well as Don Johanson's work in Ethiopia, leave absolutely no doubt about that. Under the pressure of environmental change, one branch of this proto-human line moved rapidly down the road that was to lead to us. Intelligence was at a premium in a shifting world, and so were adaptability and flexibility. A creature that could scrape a living from a variety of sources had some advantage over related species that were more dependent on one food source and had a more rigid lifestyle. We have no great argument with the palaeontologists once they are discussing the changes that occurred

(239)

after the human line had become distinct, and the reasons for those changes. But what sort of animal was it that was able not only to cope with, but also take advantage of, the changing environment?

It was almost certainly very like a modern ape in many respects, able to progress efficiently on the ground by knuckle-walking, the result of its past as a brachiator. And it came from Asia. We know that the modern apes are brachiators, modified by evolution and natural selection to enable them to swing through the trees, hanging below the branches. We also know that man, in his anatomy, bears the unmistakable stamp of brachiation. Hitherto there has been no satisfactory explanation for the puzzle of this similarity, because there are no fossil ancestors of the modern apes that are not also ancestors of man, and what few fossils there are show no evidence of being brachiators. The best that could be done was to postulate that the anatomical similarities were the end result of the process called parallel evolution, whereby the ancestors of man and the ancestors of the modern apes, though separate, led lives so similar as to equip their descendants with very similar anatomies.

The molecular evidence solves the mystery. It tells us that the five hominoid lines — gibbon, chimp, orang-utan, man and gorilla — share a common ancestor no more than 12 million years ago. That common ancestor was a fully developed tree-dweller whose brachiating lifestyle and anatomy we have inherited and make so much of in our sports and games. We are brachiators because for many millions of years our ancestors depended on brachiation to make a living, and because their adaptations served us well in our new life on the ground. The mystery brachiator who is the ancestor of all modern apes and ourselves was probably the only surviving line of a multitude of dryopithecines, one of which also gave rise to the dead end that is *Ramapithecus*. This species, so often called the earliest hominid, simply cannot be a hominid: it is too old. And yet although *Ramapithecus* is gradually losing its influence, it continues to hold many palaeo-anthropologists in its thrall.

The ancestor of the apes must have been a tree-dweller, a way station between the arboreal monkeys and the terrestrial great apes. And because the gibbons and the orang-utan, the modern tree-dwelling apes, are found in the jungles of Asia, it is safe to assume that the common ancestor of the African threesome arose in those same jungles. The Asian jungles suffered less during the climatic changes leading up to the Ice Epoch than the African jungles;

(240)

nevertheless, and despite the more gentle nature of the changes, retreating forests and spreading savannah played their part in luring some of the Asian apes down onto the ground. There was no forest connection between Asia and Africa at the end of the Miocene, some 6–8 million years ago, but that provides no block; the ancestor of the African apes could very easily have made the journey overland, knuckle-walking as the modern great apes (and, occasionally, man) still do.

The ancestors that arrived in Africa must have been fairly well adapted to life on the ground. Whether they already walked upright is impossible to say. In any case, when they arrived in Africa they were faced with new ecological opportunities. The forests still existed, though in much smaller areas since the climate had changed, and savannah was the dominant theme. As far as we know, there were no tree-dwelling apes in Africa at the time (*Ramapithecus* may have been around, but what little evidence we have suggests that it got its food on the ground) so that this niche was temptingly empty. Those members of the Asian immigrants that sought a return to a life in the trees would have encountered little resistance. Perhaps the entire population of immigrant Asian apes, in all likelihood restricted to a few pioneers, took to the trees for a few hundred thousand years, going back to that life and increasing in numbers. But the forests continued to shrink, and the very success of the species in its new niche contributed to the squeeze on the population. As in Asia before, some individuals moved out of the trees and onto the plains at the edges of the forest. Several niches were open, and the ancestral ape split to exploit alternative lifestyles. Some became chimps, others gorillas, others hominids.

The rest of the story, while not simple, is easy to agree upon. The apes stayed apes while the early hominids were propelled by the society they had invented into an ever more adaptable and intelligent lifestyle. Of course, we should stress again that it didn't really happen exactly like that. It wasn't a single resolute band of apes that trekked to Africa, just a slow progression by each generation. Nor was there any conscious will involved in the evolutionary changes. Just that among the forest-dwelling apes, the successful ones were, we assume, rather like their forebears, while among the hominids those who were more 'human' were more successful.

This account is a synthesis, bringing together dates and relationships uncovered by molecular methods, a few fossils uncovered by palaeontologists, and information about environmental changes uncovered by climatologists and geophysicists. The molecular clock tells us when particular events were taking place. It says that the common ancestor of man, chimp and gorilla existed about 4½ million years ago. Any fossil older than that cannot be the first hominid. With that in mind we have constructed a story of man's ancestry – and, incidentally, that of chimp and gorilla – that is not contradicted by any fossil evidence. We have also gone so far as to show that it is entirely possible that the direct common ancestor of ourselves and the apes was more man-like than ape-like, in that it could easily have been bipedal. Whatever the genetic changes needed to make a man instead of an ape, they are obviously tiny, just one per cent of the total genetic information, and who is to say that they couldn't have been cleanly and simply reversed, to produce a quadrupedal knuckle-walker from a bipedal ape? Quite apart from being plausible, this too is not contradicted by firm evidence, but even so, we would not want to push the speculation too hard. The information to decide one way or the other doesn't yet exist. But this overall synthesis, leaving aside the question of what the common ancestor was like, answers directly the puzzle of our shared anatomy with the apes. It makes sense of a whole host of other observations. Why, then, has it been either ignored or reviled?

Part of the problem, as we have already suggested, is man's arrogance and self-centredness. We tend to view ourselves as something special and to put ourselves above the rest of nature and one way we do that is to push our origins back in time. The molecular clock produces a birthdate for man that appears to be uncomfortably recent. We don't ourselves see why a recent origin makes most people uncomfortable, but it clearly does. We cannot understand why the idea of belonging to an ancient lineage should be so preferable. For one thing, the geological record suggests that there is an average lifetime for lineages, so that the older we are the more likely we might be to go extinct in the near future. Since many people worry about our going extinct (again wanting to put us above nature) an ancient origin for our line should increase their insecurity. Either way, what difference does it make whether our shadowy ancestor, leading its essentially primate life, was also the ancestor of the modern African apes or whether it gave rise only to ourselves? It doesn't detract from our achievements, nor excuse our errors, one bit.

But man's insecurity as expressed in his arrogance and egocentricity cannot be the whole story, for it isn't just that ordinary people around the world have failed to appreciate the molecular story. Indeed, they have not had the chance even to hear the new version, for the ones who spread the word on human beginnings are the media heroes of palaeontology, and they have been less than willing to accept the molecular evidence. This is not the way of true science; as we said, it is as if theoretical astronomers had completely ignored the discovery of pulsars. Why should palaeo-anthropologists be so different? We asked Vincent Sarich and Allan Wilson directly.

At first both of them expressed some wariness; they were still unwilling to go on the campaign trail and had no wish to emulate the media heroes of palaeontology. They prefer to let the facts speak for themselves, and be tested in the scientific community, and were initially unhappy at the prospect of a popularisation of their work. 'These people are bypassing the normal process of scientific review,' said Wilson. 'I don't think one can blame the person particularly because it's the social context that they're in. That's one reason why I cringe a little bit about your wanting to put something in a book.' We suggested to Wilson that as nobody was publicising his work, the traditionalists had the field to themselves and were winning. 'But what does winning mean?' he countered. 'It may be winning in some personal sense but in terms of the progress of science that's not winning.'[9] Wilson is interested in the progress of science, not personal glory. Nevertheless, rather than run the risk of being misinterpreted Wilson and Sarich agreed to discuss not only the work itself but also some of the peripheral issues that make the business of science so fascinating.

The first thing Sarich noted was that the phenomenon is nothing new; it goes back to the very beginning of the subject. 'Although Nuttall's work was well known and quoted by most people writing in this field from the 1920s onwards,' Sarich asserts, 'it really made very little impact on anyone's thinking.'[10] But why? 'I tried to figure out why,' he continued, as we sat having a beer after a meeting of the American Association for the Advancement of Science in Toronto. The bar was noisy and crowded but he had obviously thought this question through many times and had no need to gather his thoughts before replying. 'I think the basic blame, if one wishes to put it that way, is to the workers themselves — the molecular workers themselves. If you look in the literature describing this work, their papers, you never see the data presented in an evolutionary framework, that is in some sort of tree-like arrangement.' He fished around among his papers, brought out some of Nuttall's endless tables of numbers, and went on. 'With all the good faith in the world it is difficult to see the evolutionary message in this picture.'[11]

All that is undoubtedly true, but it belongs way back in the past. Once Morris Goodman had presented his data in a real evolutionary tree there was no longer any excuse. The implications for evolutionary ideas could hardly be ignored. And once Sarich and Wilson had published their molecular timings they certainly should not have been ignored — and they weren't, at least not totally. But they weren't taken seriously either. Critics raised objections that showed only an astounding ignorance of the procedures. For example, David Pilbeam, in his introductory textbook, has this to say: 'By making the assumptions that albumin has evolved at a constant rate in all lineages and that this rate can be measured they [Sarich and Wilson] have calculated the times of divergence of various lineages.'[12] Sarich bridles at this and other similar quotations, and with good reason. 'No, one doesn't make any assumptions about rates of change; one measures them.'[13] He went on in similar vein, explaining as we have done exactly how the properties of the clock have been tested and how the overall picture, from immunological studies, sequencing of proteins and even sequencing of the bases along the DNA, is self-consistent and mutually reinforcing. All those investigations of the molecular clock are in the scientific literature, available for all to see. Not one of the so-called scientific objections that has been raised has stood up to scrutiny, but still the detractors remain. Could that be because Sarich and Wilson plunged straight in to the primates, and man himself?

'That was probably a bad mistake,' Wilson concedes. 'We tried to do several impossible things at once. We used a technique that to most people hadn't been calibrated. The idea that it could be calibrated was considered ridiculous because everybody who asked their immunochemical neighbour "Is this possible?" was told "No, it's impossible." That's why we had to spend almost fifteen years after we'd done that work calibrating the method.'[14] Now that the method has been calibrated, and considering the remarkable consistency of the whole picture, it is surely time for the palaeontologists to come to grips with the reality of the molecular clock and to make use of the information it provides. 'I think that what the anthropologists did was to ignore or to ridicule and not to really invite us to discuss things with them. I don't think I was ever invited in all these years to give an anthropology seminar,' Wilson told us.[15]

Why were the palaeo-anthropologists so wary? Interestingly, it was shortly before Sarich and Wilson began their collaboration that David Pilbeam and Elwyn Simons, later two of the major critics of the clock, began to re-assess the various fossils now classified as *Ramapithecus*. They, more than anyone else, promoted the view that this fossil was the earliest hominid. Certainly at 14 million years old it was believed to be the earliest representative of the line to man after that line had split from the ape line. 'Whether or not *Ramapithecus* is to be described as a hominid depends on where along the line leading to man the hominid-pongid boundary is to be drawn,' Pilbeam says. 'Of course, this boundary is arbitrary,' he continues. He then proceeds to draw it: 'I would classify *Ramapithecus* as a hominid.'[16] He may have changed his mind a little now but even then the molecular evidence proved that *Ramapithecus* could not possibly be a hominid, not unless one was willing to call all the modern apes hominids too.

Sarich is convinced that this unfortunate coincidence, that Simons and Pilbeam were advancing the idea of '*Ramapithecus* is a hominid' at the same time that he and Wilson were arguing for a very recent origin for the hominid line, had a lot to do with the difficulties the molecular clock experienced. 'There was a vested interest in people like Pilbeam and Simons not seeing it [their new idea] shot down right away,' Sarich says, 'especially by this peculiar methodology that we were dealing with.' But there was more to the fossil-hunters' refusal than this. 'The palaeontologists as a community, never mind specific palaeontologists, have a vested interest in maintaining their primacy as arbiters' of how animals are related, especially how they are related in time. 'It's very difficult,' Sarich complains, 'to break through the idea that you can do temporal historic reconstructions without the palaeontologists. That is very, very hard for palaeontologists and non-palaeontologists to accept.[17] Hard it may be but accept it they must.

People like to think they are special and automatically reject any suggestion that the human species is subject to exactly the same rules as every one of the millions of other species on Earth. Are you ready to accept the chimpanzee and the gorilla as at least our equals among the intertwining branches of the evolutionary bush? If not, reflect on the great irony that a species which has got to where it is today by being adaptable and flexible — by taking advantage of, not running away from, changing circumstances — should be so conservative and blinkered in its attitude towards understanding the changing story of its own origins.

As a species we are inquisitive and we are arrogant. We are interested in ourselves and we are fascinated by our origins. We want to know where we came from, what changes made us, and where we are going. We want the solution to the monkey puzzle. The molecules say that our closest relatives are the chimpanzee and the gorilla, and that the three of us went our separate ways 4½ million years ago. They don't tell us why, and we want to know; now that we have the broad outlines of the solution we want the details too. It is up to all scientists to join together and seek the answers, for we will never get anywhere by ignoring the truth.

Our genes are, for 99 per cent of their length, the same as those of the apes. Our bodies are the bodies of infant apes. Our minds, however, which do set us a little way apart from the apes, seem unable to accept this information. We began, in the year of the centenary of his death, with Charles Darwin's opening words to *The Descent of Man*; it is fitting that we end with his ending:

'We must, however, acknowledge, as it seems to me, that man with all his noble qualities still bears in his bodily frame the indelible stamp of his lowly origin.'

EPILOGUE:
THE HUMAN PUZZLE

We may have given the impression that molecular biology has solved all the puzzles that beset man's origins, and that it is simply a question of anthropologists turning their attention to the new framework to flesh out the bones. When we began this book it seemed that way to us. But molecular evolution is still a living science, and is still capable of turning up new discoveries that have as much power to surprise as the original discoveries of Sarich and Wilson had back in 1967. One of those surprises could mean that the traditional picture, and indeed our own revised view, of mankind's origins could be very seriously in error in one important respect. Instead of arising in Africa, *Homo sapiens*, the species to which we belong, might have begun in Australia.

This astonishing idea emerges from the most recent work being done in Allan Wilson's Berkeley laboratory. As we write it is unfinished and unpublished, but it is so important that we cannot ignore it. The all-important process of peer review, whereby scientific colleagues scrutinise and test published results, has not yet had a chance to examine this work; it may prove to be mistaken, but given Berkeley's track record we think that this is rather unlikely.

The story belongs really to Becky Cann, a young graduate student working for a doctorate with Wilson and Sarich, the same one who pointed out some of the faults with Owen Lovejoy's ideas on the origin of man. She had chosen as her topic a study of the genetic material of different human groupings; but the DNA she is working on is rather unusual. Most of the DNA in every cell sits in the nucleus, controlling the operation of the cell, but there is also a certain amount of genetic material outside the nucleus. It is to this extra-nuclear DNA that the Berkeley lab's attention has switched, in particular to the DNA contained in the tiny organelles called mitochondria. These are sausage-shaped bodies that are responsible for manufacturing the energy-rich compounds used by the cell. For that reason they are often referred to as the power stations of the cell. Each mitochondrion contains a chromosome, a circular strand of DNA much smaller than a nuclear chromosome. The mitochondrial chromosome is about 16,000 bases long, compared to the billions of bases of an average nuclear chromosome, but it contains, among other things, the vital information needed to build mito-chondrial enzymes. Interestingly, mitochondria are inherited asexually. All the mitochondria in every cell of the body appear to come from the mother's side, via the egg, and none from the sperm. This means that the genes of mitochondria may never be subjected to shuffling and recombination, they simply accumulate mutations that are passed on through the female line. As we shall see, the special mode of inheritance of mitochondrial DNA is of some importance, but even more important for the study of evolutionary relationships is the fact that mitochondrial DNA evolves much faster than DNA in the nucleus.

Wes Brown and Matt George, working in Wilson's lab, have calculated that it takes about a million years for DNA in the nucleus to accumulate just two mutations in a thousand bases.[1] In the mitochondria, over the same time, twenty out of a thousand bases will change, a ten-fold increase. Overall, evolution proceeds at a pace that is dependent on two separate processes. There is the speed at which mutations occur in the DNA, often called mutation pressure, and there is the entirely independent rate at which those mutations become a permanent part of the genetic endowment of the population, the so-called rate of fixation. The first is the chance element of evolution, the second is the result of natural selection. The rate at which changes accumulate in the DNA of the nucleus is thought to be kept down by the second factor, because most changes will be harmful and so will be selected against and will not be fixed. The ten-fold higher rate of change in the mitochondria could be due either to more rapid mutations, or to the absence of selective pressure, which would enable all sorts of mutations to become fixed. The evidence favours the former view, but in any case it doesn't matter *why* mitochondria evolve quickly; what matters is that they do. 'Regardless of the reason for the high rate,' Allan Wilson explains, 'it can be of practical value to an evolutionary biologist who wants to get a magnified view of the genetic differences among individuals or species.'[2] With that magnifying glass we can look in more detail at some of the relationships we already know. More importantly, we can see relationships that are invisible with any other technique.

As far as that old staple — the split between man, chimp and gorilla — is concerned, mitochondrial DNA offers no astounding new insights. It does, however, confirm all that was known before, and with a great deal more accuracy. There are about 50 differences between the mitochondrial DNAs of man and chimp, so that if they occurred equally in both lineages there have been 25 changes in each in 4 million years, in round terms one change every 150,000 years. The method used to compare mitochondrial DNAs can easily pick up a single change, so that it can give times to within a window of 150,000 years. 'You can't do much better than that,' Sarich points out, and even at that level it is extremely difficult to separate man, chimp and gorilla.[3] Early on it looked as if mitochondrial DNA gave a slight edge to chimp and gorilla being associated, that is, sharing a common ancestor after they had diverged from our line. 'In spite of a monumental study by Steve Ferris and Ellen Prager, it's still frustratingly difficult to decide whether that's a three-way split or not,' Wilson told us. 'Chimp-gorilla association still has a slight edge, but percentage wise it's decreasing as we get more sequencing done.'[4] So the mitochondria confirm the details of the man-ape split. They also agree with the other techniques on the times of the various splits. Wilson points out that the mitochondria provide a new and independent molecular clock: 'We get the impression that mitochondrial DNA's divergence is mirroring the time, the history of . . . populations in much the same kind of way that we thought proteins were, that one has a fairly relentless accumulation of point mutations with time.'[5] The difference is that this clock is ten times more precise than any other, although it is restricted to the relatively recent past.

Normally it is very hard, if not impossible, to tell when two individual members of a single breeding population split from one another, because the genes in the nucleus are forever being shuffled and mixed as a result of sex and reproduction. Differences could be due to clock-like mutations, but they might just as easily not be. Mitochondria, because they are inherited solely from the mother, provide a window on to individual splits. And because the DNA in mitochondria evolves so fast, it is possible to determine with greater accuracy when two mitochondrial genes were last present in the same female body. Comparisons of individuals from several species reveal a very strange thing. The individuals of an average mammal species differ by about 1.5 per cent in their mitochondrial DNA, much higher than the difference seen in nuclear DNA. But two humans, plucked at random from a crowd, differ far less, by only 0.4

per cent. Chimps and orang-utans, in this respect, are like other mammals. Two chimps chosen at random are likely to differ by 1.5 per cent. As Wilson puts it, 'the human species is quite unusual in having a very low level of variation. Among natural species I don't know of any other cases like this.'[6] At last, we've found something biochemically unusual about *Homo sapiens*, and it leads us to the surprise we talked about earlier.

Fewer differences implies less time for those differences to accumulate. Human mitochondrial DNA is about six times less divergent than that of other mammals. That means that the modern human species − *Homo sapiens* proper − arose very much more recently than the average mammalian species. The members of a small group that begins a species are closely related to one another and individuals in the group share a great deal of their mitochondrial genome; only after the new species proves a success and begins to spread do the mitochondrial lineages begin to split and accumulate their own individual characteristics; in the case of woman, they haven't been doing so for long. The average living species is about a million years old, that much has been calculated from the fossil record and is confirmed by the molecular clocks. The modern forms of chimpanzee and orang-utan are certainly at least as ancient, possibly more so, and their mitochondrial DNA is like other mammals'. But the fossil evidence for man, truly modern *Homo sapiens*, goes back only to 39,000 years ago, and this is reflected in our mitochondrial DNA.

So that is one thing to bear in mind, that truly modern man is very young indeed. What now of the differences between groups of humans? We have said that, as far as humans are concerned, the strictly biological concept of race has no meaning. But there are obvious differences between human beings, and we can easily divide Old World people into four major groupings: Blacks, Whites, Australians and Orientals. A close look at blood proteins reveals a few differences among these four groups and allows one to construct an evolutionary tree. According to this protein tree, the Australians and Orientals diverged about 40,000 years ago, but the major grouping of mankind arose about 100,000 years ago with a three-way split into Blacks, Whites and the Australian-Oriental lineage. This date, while a little further back than many anthropologists would accept, is not inconsistent with what we know of the transition between *Homo erectus* and *Homo sapiens*, which began perhaps 500,000 years ago. It suggests that the present subdivision of mankind arose after mankind had become anatomically modern and sapient, and that there was but a single lineage leading from *Homo erectus* to *Homo sapiens*.

Becky Cann comes up with a completely different family tree. She has examined in great detail the mitochondrial DNA of more than a hundred human beings, and she gets a very peculiar set of results. For the most part Blacks, Whites, Australians and Orientals are very similar to each other, differing by about one in every 200 bases. But some individual Orientals have mitochondrial DNA that differs a little from the main group, and some Australians are very different indeed from the main bunch of people. It is as if the vast majority of people form a single tight grouping, which includes Blacks, Whites, Orientals and Australians, but that there are tiny offshoots from this group, one containing some Australians and the other some Orientals, and the people inside these offshoots, while different from the main group, are similar to one another. As if the odd branching order weren't enough, mitochondrial DNA also puts puzzling times on the various splits; it says that the discrepant Australians split from the main group about 400,000 years ago, that the divergent Orientals originated 100,000 years ago, and that the other groupings are about 40,000 years old.

These timings really are quite a shock, and it is very hard to know how to respond to them. Part of the problem is that information of this sort has a long history of misuse and abuse. In the 19th century, after the theory of natural selection had been assimilated, well-meaning scientific racists attempted to justify current practices in society by arranging the races of man into an ascending order of evolution. Those races that had been evolving longest were, of course, the most advanced, and naturally the European scientists who drew up these ladders were certain that their race had been human longest. Negroes and Orientals were not so advanced, and therefore it was permissible to treat them as not so human. Stephen Jay Gould sums up this approach as follows.

'Clearly, science did not influence racial attitudes in this case. Quite the reverse: an a priori belief in black inferiority determined the biased selection of "evidence". From a rich body of data that could support almost any racial assertion, scientists selected facts that would yield their favored conclusion according to theories

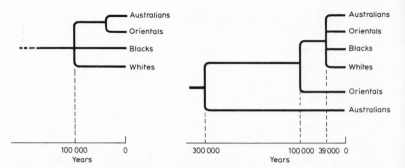

Left, proteins and DNA in the chromosomes provide a picture of human evolution that times the origin of human races to about 100,000 years ago, close to the onset of the most recent glaciation. Right, mitochondrial DNA puts the origins of race and our species much further back, and hints that the initial transition to modern man could have occurred in Australia, not Africa.

currently in vogue. There is, I believe, a general message in this sad tale. There is not now and there never has been any unambiguous evidence for genetic determination of traits that tempt us to make racist distinctions (differences between races in average values for brain size, intelligence, moral discernment, and so on). Yet this lack of evidence has not forestalled the expression of scientific opinion. We must therefore conclude that this expression is a political rather than a scientific act — and that scientists tend to behave in a conservative way by providing "objectivity" for what society at large wants to hear.'[7]

The history of anthropology is replete with scientific racism, and the news that the Australians might represent the oldest human population brings dark memories to the fore and stirs old arguments. 'I think the majority of anthropologists are tinged by racism,' Allan Wilson said to us in conversation, 'as most of us are, and that their science hasn't been a quantitative one and an objective one.' These data from Berkeley are quantitative and objective, and Wilson confesses he is 'very worried about how to present this in a responsible way'.[8]

Leaving aside, for a moment, the implications of this work, how can we interpret the discrepancy between the simple pattern offered by blood groups and proteins and the altogether more baffling tree revealed by the mitochondria? 'One way,' Cann and Wilson explain, 'is to say that *Homo sapiens* arose somewhere in the Old World, and as they expanded out and wiped out *Homo erectus* as they went, the last place they would get to was Australia. The population that entered Australia from nearby parts of Indonesia may have been a mixed one, the result of contact between an incoming *sapiens* population and a resident population that arose independently from *Homo erectus*. In this case, the nuclear genomes of the two populations would be blended and assorted, but the mitochondrial genomes would remain separate, so that you would find representatives of the two lineages present in Australia.'[9] In effect, this argument runs, the modern Australians are hybrids between *Homo sapiens* and another people, derived from *Homo erectus*; blood proteins don't show this because they stem from nuclear DNA that has been thoroughly mixed, but the mitochondrial DNA, being passed on only by females, reveals the presence of females from two very distinct lineages in Australia, and the early date for these Australians reflects not the time of the invasion of *sapiens* but the split of the pre-sapient Australians from their ancestors. Exactly the same considerations apply to the slightly divergent Orientals, with waves of spreading of the newly evolved *sapiens* and occasional incorporation of females into the *sapiens* breeding population.

'It's kind of a touchy business to know how to deal with,' Wilson told us. 'The view that I've presented is the one that Europeans like, that Caucasians are the people who did the spreading. There's an alternative, which is that *Homo sapiens* originated in Australia and the spread occurred in the other direction.' It was late at night and we were talking in the bar of King's College in Cambridge, where we received this possibility with what must have seemed a little too much scepticism. Wilson sighed. 'Most people would consider that so unreasonable that it would prove how crazy we are.'[10] But he went on to persuade us.

We know that *Homo erectus* was present throughout the Old World for the best part of 2 million years. And we know that the oldest fossil of truly modern man, which is just 39,000 years old, comes not from Europe but from Indonesia. 'So,' says Wilson, 'what's the possibility — I mean, it seems too far out I admit, but I want to explore it more — that while *erectus* was muddling along in the rest of the world a few *erectus* had got to Australia and did something dramatically different — maybe not with stone tools —

and that that's where *sapiens* evolved and then got back to the rest of the world.'[11] By this account, the ancient timing from the mitochondria would be genuine, *Homo sapiens* would have evolved, free from competition, out of a small band of *Homo erectus* that, 400,000 years ago, somehow made its way across the Timor straits and into Australia.

And it might make sense of some very strange results from Raoul Benveniste of the National Institutes of Health. He has taken a close look at a sequence of bases along a stretch of the DNA that represent a very odd sort of virus. This virus can insinuate itself into the DNA and be passed down the generations harmlessly, and then suddenly cause cancer. The primates of Africa — baboons, gorillas, chimpanzees, vervets, and so on — share very similar sequences for this region, suggesting that the virus is a common hazard that they have faced for a long time. But man's sequence is not like that of the African primates, and is more like that of the gibbon and orang-utan. Benveniste concludes that 'most of man's evolution has occurred outside Africa'.[12]

It is an astounding thought, that Australia may be the cradle of mankind. And there is a little evidence in favour of the alternative suggestion, that Australians are of hybrid origin. It comes from the work of James Neel, of the department of human genetics at the University of Michigan medical school, one of the most eminent American geneticists.[13] Neel is interested in mutation rates of human populations, and he has investigated the people of London, and Ann Arbor in Michigan, as well as Japanese living in the rebuilt cities of Hiroshima and Nagasaki. With all the modern techniques at his disposal, he has been unable to find any evidence of an increased rate of mutation in the A-bombed cities, and indeed it is questionable whether he can detect any new mutations at all in people by comparing new-born babies with their parents. But there is another approach, and that is to study what Neel calls rare variants or private polymorphisms. These are genes that occur at very low levels in the population, below one per cent, and are often restricted to discrete geographical locations. Genetic variability is called a polymorphism if the frequency is quite high, like the polymorphism for blood group, or eye colour, where a sizeable proportion of the population carries each variety. Rare variants could be mutations that arose in one part of the world and haven't spread much, so they might, according to Neel's reasoning, provide an insight into the rate of mutation.

Londoners, for example, are a very mixed bunch. If you could trace their ancestry back you would find that they come from many different tribes, or breeding groups, probably more than 100 of them. Now if each of those tribes had some little local mutation — a rare allele — that it has carried with it, the people of London, taken together, ought to have lots of different rare variants. And indeed, when you look at the people of London and other cosmopolitan cities, as Neel has done, you find that there are quite a few rare alleles to be discovered; on average one in a thousand people carries at least one of these variants. In addition to these 'civilised' people, Neel also looked at what he called 'tribal' people, mostly from Australia and New Guinea but also the Yanomamo Indians of the Amazon basin. These people should have very few rare variants, because they have not bred widely with other tribes. But, the tribal people of Australia and New Guinea have ten times more rare variants than the civilised people of London, Ann Arbor, Hiroshima, and Nagasaki. Eleven out of a thousand have a rare variant. Neel's interpretation of this is that the mutation rate is higher in tribal people, perhaps as a result of disease but also probably because of the foods they eat. 'Molds may be more common in the tropics, ' Neel says, and many moulds are known to cause mutations. 'Meat is often cooked by direct contact with the fire, in the process acquiring a thick char,'[14] and there is growing evidence that meat burned in this way can also cause mutations. And lots of the plant foods contain protective compounds that may tend to increase the number of mutations.

Allan Wilson sees these results differently. 'Neel concluded that tribal populations have high mutation rates and that we have low mutation rates, which could lead others to conclude that there's something good about the civilised life, even about bombs,' he explained to a seminar at the Imperial Cancer Research Fund in North London, a group with more than a passing interest in Neel's discoveries. 'But,' he continued, 'this result is, I think, exactly what you would expect if the Australian population was of hybrid origin. The old population, that got swamped by the incoming population, had polymorphisms, but now they've been diluted into the "rare allele" category.'[15] In conversation he expanded on this alternative view of Neel's results. 'This explanation for the high incidence of rare alleles in Australia is the one that Becky [Cann] favours. It is . . . a hybrid population where there's a minority of alleles that stem from this ancient population that got recently engulfed. And those are things that were major alleles, they're not new mutations. I'm just delighted to see, not that Neel's made an error, but that he's summarised these data which fit so well with what we would predict from the mitochondrial hypothesis. I just see it as a nice fulfilment of the expectations that we would have had.'[16]

If this view is correct, the mitochondria of the offshoot Australians represent the mitochondria of their late-*erectus* type ancestors, passed down through the female line. But the rare variants, or alleles, in their nuclear DNA are the polymorphisms of their ancestors diluted by the DNA of the invading sapients. So, what happened in Europe and Asia during the great spread of *Homo sapiens*? Wilson believes that if you had looked at the people of the Old World about 50,000 years ago you would have found a mixture very similar to that of modern Australia. Archaic people, like the Neanderthals of Europe or the Solo types of Indonesia, would have been swallowed up into the *sapiens* populations, creating the combination of old and new mitochondrial lineages in one species. This idea, that modern man interbred with local populations, is not terribly new, but has never had any hard evidence to support it. Mitochondrial studies not only offer evidence that this may have been the way things happened, but could also provide an answer to the problem of what became of Neanderthal man. There is as yet no evidence of any offshoot among the Whites whose mitochondria Becky Cann has looked at, but perhaps, lurking within a very few Europeans, there are the mitochondrial genes of Neanderthal woman, who was incorporated into their tribes by the advancing sapient hordes. The question that remains is where would one look for people with Neanderthal mitochondria? We might guess that somewhere out of the way, far from mainland Europe, might be the place; perhaps Lapland would be a good place to start.

Hybridisation between invading *sapiens* and existing pre-*sapiens* people is one way to interpret the mitochondrial data. But what of the alternative, that Australia, not Africa, is the cradle of mankind? It's an idea that, according to Wilson, 'physical anthropologists and many other people would consider unreasonable.'[17] That, however, is no reason not to take it seriously. It is interesting that the oldest modern human fossils come from Indonesia, very close to Australia, but aside from this there is not a great deal of evidence. The point is that there was not a great deal of evidence for an African origin until people began to look, and people haven't really looked in Australia. 'Just think of the resistance that the Leakeys and people encountered to the idea that humans arose in Africa,' Wilson points out. 'Now it's considered OK, but European scientists didn't think much of the idea for a long time. We don't have evidence of humans in Australia before 40,000 years, but Australia's a terrible place in terms of sedimentary rocks, and there's almost no volcanic activity, so you don't have well-dated layers in which you can look for fossils.'[18] That is very true, and the thick, easily dated beds of East Africa have become the focus of attention partly because they are relatively easy to work, rather than because they really hold the key to the emergence of modern man. But there is another problem that Wilson puts his finger on: 'What would you look for anyway? It may be that it wasn't a stone tool type culture.'[19]

So, after more than a decade of essentially confirmatory work, molecular studies of evolution have once again produced an enormous surprise, that some Australians and some Orientals are very different mitochondrially from other people. And they have suggested that these groups go back possibly to 400,000 and 100,000 years ago. There are two possible explanations; either *Homo sapiens* invaded Australia from elsewhere and mingled with another group of people there, whose genes remain in the present population as rare alleles, or *Homo sapiens* arose in Australia and spread, very late, to cover the Old and New Worlds, probably in more than one wave. The evidence for the two possibilities is not equally matched; Cann and Wilson favour the former possibility. But Wilson is wary of pushing the hybridisation hypothesis, because of the misuse he fears will be made of this information in subjugating native Australians — to which one can only reply that the subjugation has not been held back through lack of this sort of knowledge. In any event, Wilson is 'tempted to think about the other idea more seriously'. The problem, as he well knows, is that the idea that Australia is the cradle is, as he puts it, 'one that the archaeologists and the anthropologists will be able to ridicule rather easily'. The prospect of being held up to ridicule might worry some scientists, but not Wilson. He says, with humility, not arrogance, 'I know from previous experience that they can feel quite confident that they're right about something when in fact they're wrong.'

Although there is some archaeological basis for preferring one hypothesis over the other, both need to be examined carefully with an open mind, searching for the clues that will tell us whether man eventually reached Australia in the course of his spreading, or whether he sprang from it and then spread across the Earth. 'I think our results should provide a stimulus to people,' Wilson says. With luck that hope, unlike Nuttall's in 1904, will be realised, but the probable outcome of further investigations is unclear. What is clear is that the information hidden in the DNA of mitochondria has set us another puzzle to untangle.

ACKNOWLEDGEMENTS

Thanks are due to those who gave permission to use material quoted in the text, from published and unpublished sources and from interviews, all of which are noted in full in the References that follow.

The black and white photographs have been used with the permission of the following: page 1 (*left*) Allsport Photographic/Tony Duffy, (*right*) Bruce Coleman/Helmut Albrecht; page 2 (*top*) Bruce Coleman, (*bottom*) Bruce Coleman/Helmut Albrecht; page 3, Blackfriars Settlement; pages 4, 5, 6, 7, 8 and 9, John Reader; page 10, Ardea London/Clem Haagner; page 11, Ardea London/(*top left*) Ian Beams, (*top right*) John Clegg, (*bottom*) Ken Hoy; page 12 (*top and bottom*) Bruce Coleman/Jane Burton; page 13 (*left*) J. Yunis, (*right*) M. Goodman; page 14 (*top*) Ardea London/Jean-Paul Ferrero, (*middle*) Ardea London/Francois Gohier, (*bottom*) Bruce Coleman/Al Giddings; page 15, Ardea London; page 16, Alan Hutchinson, Orion Press, BBC Picture Publicity and Stephen Benson.

The drawings in the text are by Neil Hyslop.
Especial thanks to Andrea Moore for picture research.

REFERENCES

Chapter One

[1] Sherwood L. Washburn, 'The evolution of man', *Scientific American*, 239(3): 194–208, 1978. p 204.

[2] Virginia Avis, 'Brachiation: the crucial issue for man's ancestry', *Southwestern Journal of Anthropology*, 18: 119–148, 1962.

[3] As 2, p 145.

[4] Sherwood Washburn, 'Behavior and Human Evolution', in *Classification and human evolution*, ed S. L. Washburn, Aldine: Chicago, 1963. p 194.

[5] George H. F. Nuttall, *Blood Immunity and Blood Relationships*, Macmillan: New York, 1904. p 4.

[6] As 5, p 411.

[7] As 5, p 63.

[8] George H. F. Nuttall, 'The new biological test for blood − its value in legal medicine and in relation to zoological classification', *Journal of Tropical Medicine*, 4: 405–408, 1901. p 405.

[9] As 5, p 410.

[10] As 5, p 4.

[11] Morris Goodman, 'Evolution of the immunologic species specificity of human serum proteins', *Human Biology*, 34: 104–150, 1962. p 145.

[12] As 11, p 148.

[13] Morris Goodman, 'Serological analysis of the systematics of recent hominoids', *Human Biology*, 35: 377–424, 1963. p 400.

[14] As 13, p 399.

[15] Vincent Sarich, interview with the authors, 3 February 1981, London.

[16] Vincent Sarich & Allan Wilson, 'An immunological timescale for hominid evolution', *Science*, 158: 1200–1203, 1967. p 1202.

[17] As 16, p 1202.

[18] Marie-Claire King & A. C. Wilson, 'Evolution at two levels in humans and chimpanzees', *Science*, 188: 107–116, 1975.

Chapter Two

[1] E. J. Dupraw, *DNA and Chromosomes*, Holt, Rinehart and Winston: New York, 1970.

[2] For an outstanding history and explanation of molecular biology see Horace Freeland Judson, *The Eighth Day of Creation,* Simon & Schuster: New York, 1979.

[3] Fred N. White, 'Respiration', in *Animal Physiology: Principles and Adaptations,* ed. M. S. Gordon, Macmillan: New York, 1977.

[4] R. D. Martin, 'Review', *Primate Eye*, No. 16: 32–33, 1981. p 32.

[5] Quoted in Roger Lewin, 'Evolutionary theory under fire', *Science*, 210: 883–887, 1980. p 883.

[6] Steven M. Stanley, *Macroevolution*, Freeman: San Francisco, 1979. p 30.

[7] T. R. Kemp, interview with JC, 9 April 1981, Oxford.

[8] Donald C. Johanson & Maitland A. Edey, *Lucy: The Beginnings of Humankind,* Simon & Schuster: New York, 1981.

[9] Richard E. Leakey, *The Making of Mankind,* E. P. Dutton: New York, 1981.

[10] JG was present, Cambridge, 1978.

[11] Niles Eldredge & Stephen Jay Gould, 'Punctuated equilibria: an

alternative to phyletic gradualism', in *Models in Paleobiology*, ed. T. J. M. Schopf, Freeman Cooper: San Francisco, 1972.

[12] As 5, p 884.

[13] As 5, p 883.

[14] In episode 1 of Richard Leakey's BBC television series *The Making of Mankind,* and *The Listener* transcript. A paper is due to appear in *Nature* soon.

[15] Cited by Pere Alberch, S. J. Gould, G. F. Oster & David B. Wake, 'Size and shape in ontogeny and phylogeny', *Paleobiology*, 5: 296–317, 1979.

[16] As 5, p 886.

[17] Jorge J. Yunis, Jeffrey R. Sawyer & Kelly Dunham, 'The striking resemblances of high-resolution G-banded chromosomes of man and chimpanzee', *Science*, 208: 1145–1148, 1980.

[18] Quoted by Stanley (ref. 6) p 6.

[19] As 5, p 884.

[20] As 6, p 81.

[21] As 6, p 82.

[22] Quoted in John Reader, *Missing Links,* Little, Brown: Boston, 1981. p 29.

[23] As 22, p 30.

[24] As 22, p 17.

Chapter Three

[1] John McPhee, *Basin and Range,* Farrar, Straus & Giroux: New York, 1981. p 126.

[2] Elwyn Simons, 'The fossil record of primate phylogeny', in *Molecular Anthropology*, eds. M. Goodman & R. E. Tashian, Plenum: New York, 1976. p 38.

[3] J. E. Cronin, N. T. Boaz, C. B. Stringer & Y. Rak, 'Tempo and mode in hominid evolution', *Nature*, 292: 113–122, 1981.

[4] David Pilbeam, *The Ascent of Man: An Introduction to Human Evolution,* Macmillan: New York, 1972. p 31.

[5] As 4, p 46.

[6] Richard Leakey & Roger Lewin, *Origins,* E. P. Dutton: New York, 1978. p 81

[7] Richard Leakey & Roger Lewin, *People of the Lake,* Anchor Press/Doubleday: New York, 1978. p 254.

Chapter Four

[1] Vincent Sarich, interview with the authors, 3 February 1981, London.

[2] Allan Wilson, interview with JC, 21 June 1981, Cambridge.

[3] Quoted in Horace Freeland Judson, *The Eighth Day of Creation,* Simon & Schuster: New York, 1979. p 305.

[4] Emile Zuckerkandl, 'Perspectives in molecular anthropology', in *Classification and Human Evolution,* ed. S. L. Washburn, Aldine: Chicago, 1963. p 246.

[5] As 4, p 247.

[6] G. William Moore, 'Proof for the maximum parsimony ('Red King') algorithm', in *Molecular Anthropology,* eds. M. Goodman & R. E. Tashian, Plenum: New York, 1976. p 120.

[7] As 6, p 117.

[8] As 2.

[9] L. B. Halstead, *The Evolution of the Mammals,* Hippocrene Books: New York, 1981. p 120.

[10] Vincent M. Sarich, 'Pinniped origins and the rate of evolution of carnivore albumins', *Systematic Zoology,* 18: 286–295, 1969. p 288.

[11] As 10, p 290.

[12] As 10, p 291.

[13] Vincent M. Sarich, 'The giant panda is a bear', *Nature*, 245: 218–220, 1973.

[14] As 10, p 291.

Chapter Five

[1] Vincent M. Sarich & Allan C. Wilson, 'An immunological time-scale for hominid evolution', *Science*, 158: 1200–1203, 1967. p 1200.

[2] As 1, p 1201.

[3] As 1, p 1202.

[4] As 1, p 1202.

[5] Raoul E. Benveniste & George J. Todaro, 'Evolution of type C viral genes: evidence for an Asian origin of man', *Nature*, 261: 101–108, 1976.

[6] Vincent M. Sarich, unpublished manuscript, p 10.

[7] Marie-Claire King & A. C. Wilson, 'Evolution at two levels in humans and chimpanzees', *Science*, 188: 107–116, 1975.

[8] Jorge J. Yunis, Jeffrey R. Sawyer & Kelly Dunham, 'The striking resemblances of high-resolution G-banded chromosomes of man and chimpanzee', *Science*, 208: 1145–1148, 1980.

[9] S. D. Ferris, A. C. Wilson & W. M. Brown, 'Evolutionary tree for apes and humans based on cleavage maps of mitochondrial DNA', *Proceedings of the National Academy of Sciences*, 78: 2432–2436, 1981.

[10] Lorraine M. Cherry, Susan M. Case & Allan C. Wilson, 'Frog perspective on the morphological difference between humans and chimpanzees', *Science*, 200: 209–211, 1978.

[11] As 6, p 6.

[12] As 6, p 6.

[13] W. O. Weigle, 'Immunological properties of the crossreactions between anti BSA and heterologous albumins', *Journal of Immunology*, 87: 559, 1961.

[14] Vincent Sarich, interview with the authors, 3 February 1981, London.

[15] Quoted in 'Molecular evolution: a quantifiable contribution', *Mosaic*, 10(2): 14–22, 1979. p 19.

[16] As 15, p 17.

[17] As 15, p 19.

[18] As 6, p 18.

[19] As 15, p 22.

[20] Vincent M. Sarich & John E. Cronin, 'Molecular systematics of the primates', in *Molecular Anthropology*, eds. M. Goodman & R. E. Tashian, Plenum: New York, 1976. p 156.

[21] Quoted in Carl Sagan, *Dragons of Eden,* Random House: New York, 1977.

[22] As 14.

[23] As 8, p 1148.

[24] Quoted in J. W. T. Moody, *Journal of the Society for the Bibliography of Natural History*, 5: 474–476, 1968, and elsewhere.

[25] Journal of the Proceedings of the Linnean Society, *Zoology* (IV): viii–xx, 1860.

[26] As 24, p 475.

[27] As 25, p ix.

[28] *Dictionary of National Biography*, Oxford University Press: Oxford, 1975.

[29] Charles Darwin, *The Origin of Species,* Penguin: Harmondsworth and New York, 1968. p 458, and elsewhere.

Chapter Six

[1] Donald C. Johanson & Maitland A. Edey, *Lucy: The Beginnings of Humankind,* Simon & Schuster: New York, 1981. p 376.

Chapter Seven

[1] Quoted in Stephen Jay Gould, *Ever Since Darwin,* W. W. Norton: New York, 1979.

[2] Robert Ardrey, *African Genesis,* Dell: New York, 1966, and *The Territorial Imperative,* Dell: New York, 1968.

[3] Quoted in Richard Leakey & Roger Lewin, *People of the Lake,* Anchor Press/Doubleday: New York, 1978. p 262.

[4] Alister Hardy, 'Was man more aquatic in the past?', *New Scientist,* 7: 642–645, 1960.

[5] Elaine Morgan, *The Descent of Woman,* Stein & Day: New York, 1972. A recent addition to this literature, which stresses the female role in evolution but unaccountably ignores Morgan and Hardy, is Nancy Makepeace Tanner's *On Becoming Human,* Cambridge University Press: Cambridge and New York, 1981.

[6] Fred Hoyle & Chandra Wickramasinghe, *Diseases From Space,* Harper & Row: New York, 1980.

[7] C. Owen Lovejoy, 'The origin of man', *Science,* 211: 341–350, 1981.

[8] Rebecca L. Cann & Allan C. Wilson, 'Female role in hominid evolution', *Science,* in the press. (ms. p 4.)

[9] For details, see Steven Weinberg, *The First three Minutes,* Basic Books: New York, 1976, and John Gribbin, *Genesis,* Delacorte Press: New York, 1981.

[10] Aldous Huxley, *After Many a Summer Dies the Swan,* Harper & Row: New York, 1965.

[11] Donald C. Johanson & Maitland A. Edey, *Lucy: The Beginnings of Humankind,* Simon & Schuster: New York, 1981. p 277.

[12] As 11, p 278.

[13] As 11, p 285.

[14] As 11, p 360.

[15] As 11, p 358.

[16] As 11, p 277.

[17] *Oxford Dictionary of Quotations,* 2nd edn., Oxford University Press: Oxford, 1980. p 192.

[18] As 11, p 351.

[19] Quoted in Richard E. Leakey, *The Making of Mankind,* E. P. Dutton: New York, 1980. pp 131–132.

[20] As 19, p 132.

Chapter Eight

[1] For details see John Imbrie & Katherine Palmer Imbrie, *Ice Ages: Solving the Mystery,* Enslow Publications: Hillside, N.J., 1979.

[2] As 1, pp 19–20.

[3] As 1, p 33.

[4] Hubert Lamb, *Climate: Present, Past and Future,* vol. 2, Methuen: London, 1977. p 309.

[5] For details see John Gribbin, *Genesis,* Delacorte Press: New York, 1981, and Ursula B. Marvin, *Continental Drift: the Evolution of a Concept,* Smithsonian Institution: Washington, 1973.

[6] N. J. Shackleton & N. D. Opdyke, 'Oxygen isotope and palaeomagnetic stratigraphy of equatorial Pacific core V28-238: oxygen

isotope temperatures and ice volumes on a 10^5-year and 10^6-year scale', *Quaternary Research*, 3: 39-55, 1973, and 'Oxygen isotope and palaeomagnetic stratigraphy of Pacific core V28-239, late Pliocene to latest Pleistocene', in *Geological Society of America Memorandum No. 145*, eds. J. D. Hays & R. M. Cline, 1976.

[7] J. D. Hays, John Imbrie & N. Shackleton, 'Variations in the Earth's orbit: pacemaker of the ice age', *Science,* 194: 1121-1132, 1976.

[8] As 1, p 80.

[9] As 6.

[10] William Ruddiman & Andrew McIntyre, 'Oceanic mechanisms for amplification of the 23,000-year ice-volume cycle', *Science, 212: 617-627*, 1981.

[11] Quoted in Richard E. Leakey, *The Making of Mankind*, E. P. Dutton: New York, 1981. p 148.

[12] C.G. Turner II & J. Bird, 'Dentition of Chilean paleo-Indians and peopling of the Americas', *Science,* 212: 1053-1055, 1981.

Chapter Nine

[1] T.A. Sebeok & J. Umiker-Sebeok (eds.), *Speaking of Apes: A Critical Anthology of Two-Way Communication with Man*, Plenum: New York, 1980.

[2] R. Passingham, *The Human Primate*, Freeman: London, in the press.

[3] There is an excellent discussion of the philosophical debate and practical implications in Adrian Desmond, *The Ape's Reflexion,* Dial Press: New York, 1979.

[4] As 1, p 392.

[5] As 3, p 65.

[6] As 3, pp 102-103.

[7] Herbert S. Terrace, *Nim: A Chimpanzee Who Learned Sign Language,* Alfred A. Knopf: New York, 1979. p 22.

[8] As 7, p 4.

[9] As 7, p 20.

[10] Emil Menzel, 'Natural language of young chimpanzees', *New Scientist,* 65: 127–130, 1975. p 127.

[11] As 10, p 130.

[12] Jane van Lawick-Goodall, *In the Shadow of Man,* Houghton Mifflin: Boston, 1981.

[13] Quoted in Richard E. Leakey, *The Making of Mankind,* E. P. Dutton: New York, 1981. p 140.

[14] Quoted in Jeremy Cherfas, 'Voices in the wilderness', *New Scientist,* 86: 303–306, 1980. p 306.

[15] As 3, p 152.

[16] Douglas Gillan, interview with JC, 25 March 1981, Berlin.

[17] As 16.

[18] As 16.

[19] E. Sue Savage-Rumbaugh, D. M. Rumbaugh, S. T. Smith & J. Lawson, 'Reference: the linguistic essential', *Science,* 210: 922–925, 1980.

[20] As 19, p 923.

[21] As 3, p 148.

[22] R. V. S. Wright, 'Imitative learning of a flaked stone technology — the case of an orangutan', *Mankind,* 8, 1972.

[23] As 16.

[24] Geza Teleki, 'Chimpanzee subsistence technology: materials and skills', *Journal of Human Evolution,* 3: 575–594, 1974. p 588.

Chapter Ten

[1] Christine Janis, interview with JC, 22 June 1981, Cambridge.

[2] John Reader, *Missing Links,* Little, Brown: Boston, 1981.

[3] Allan Wilson, interview with JC, 21 June 1981, Cambridge.

[4] As 3.

[5] Vincent Sarich, 'Molecular contributions to the understanding of hominid evolution', paper to the American Association for the Advancement of Science, 7 January 1981, Toronto.

[6] As 3.

[7] Vincent Sarich, unpublished manuscript, p 10.

[8] Vincent Sarich, interview with JC, 7 January 1981, Toronto.

[9] As 3.

[10] As 5.

[11] As 8.

[12] David Pilbeam, *The Ascent of Man: An Introduction to Human Evolution,* Macmillan: New York, 1972. pp 46–47

[13] As 5.

[14] As 3.

[15] As 3.

[16] As 12, pp 95–96.

[17] As 8.

Epilogue

[1] Wesley M. Brown, Matthew George Jr. & Allan C. Wilson, 'Rapid evolution of animal mitochondrial DNA', *Proceedings of the National Academy of Sciences,* 76: 1967–1971, 1979; and S. D. Ferris, A. C. Wilson & W. M. Brown, 'Evolutionary tree for apes and humans based on cleavage maps of mitochondrial DNA', *Proceedings of the National Academy of Sciences,* 78: 2432–2436, 1981.

[2] Allan Wilson, 'The origin of genetic diversity in *Homo sapiens*', seminar to the Imperial Cancer Research Fund, 9 July 1981, Mill Hill.

[3] Vincent Sarich, 'Molecular contributions to the understanding of hominid evolution', paper to the American Association for the Advancement of Science, 7 January 1981, Toronto.

[4] Allan Wilson, interview with JC, 21 June 1981, Cambridge.

[5] As 2.

[6] As 2.

[7] Stephen Jay Gould, *Ever Since Darwin,* W. W. Norton: New York, 1979.

[8] As 4.

[9] As 2.

[10] As 4.

[11] As 4.

[12] Raoul E. Benveniste & George J. Todaro, 'Evolution of type C viral genes: evidence for an Asian origin of man', *Nature,* 261: 101–108, 1976.

[13] James V. Neel & Edward Rothman, 'Is there a difference among human populations in the rate with which mutation produces electrophoretic variants?', *Proceedings of the National Academy of Sciences,* 78: 3108–3112, 1981. p 3112.

[14] As 13.

REFERENCES

[15] As 2.

[16] As 4.

[17] As 2.

[18] As 2.

[19] As 2.

INDEX

INDEX

John Gribbin was born in England in 1946 and educated at the University of Cambridge, where he obtained his Ph.D. in astrophysics. He subsequently worked for five years on the journal *Nature,* and in 1974 he was awarded Britain's premier science-writing award, the National Award, sponsored by Glaxo, for his work on the nature of X-ray stars and quasars. In 1974 he joined the Science Policy Research Unit of the University of Sussex, and since 1978 he has been physics consultant to the periodical *New Scientist.* He has published seventeen books to date.

Jeremy Cherfas was born in England in 1951 and educated at the University of Cambridge, where he received his Ph.D. in animal behavior. After post-doctoral research in Toronto, he returned in 1978 to London, where he worked as biology editor of *New Scientist.* He is currently life sciences consultant to that periodical, and presenter of the BBC World Service's science program "Discovery," and a professor in the Zoology Department of Oxford University. He has published three other books to date.